# Laboratory Anatomy of the
# Mink

Second Edition
# Laboratory Anatomy of the
# Mink

**David Klingener**
University of Massachusetts, Amherst

QL
813
.M56
K56
1979

**wcb**
**Wm. C. Brown**
Company Publishers
Dubuque, Iowa

**Booth Laboratory Anatomy Series**

**Consulting Editors**
Ernest S. Booth
Robert B. Chiasson

Laboratory Anatomy of the Cat
Ernest S. Booth
Robert B. Chiasson

Laboratory Anatomy of the Domestic Chicken
Michael C. Robinson

Laboratory Anatomy of the Elementary Chordates
Robert B. Chiasson

Laboratory Anatomy of the Frog
Raymond A. Underhill

Laboratory Anatomy of the Human Body
Bernard B. Butterworth

Laboratory Anatomy of the Iguana
Jonathan C. Oldham, Hobart M. Smith

Laboratory Anatomy of the Mink
David Klingener

Laboratory Anatomy of Necturus
Robert B. Chiasson

Laboratory Anatomy of the Perch
Robert B. Chiasson

Laboratory Anatomy of the Fetal Pig
Theron O. Odlaug

Laboratory Anatomy of the Pigeon
Robert B. Chiasson

Laboratory Anatomy of the Rabbit
Charles A. McLaughlin
Robert B. Chiasson

Laboratory Anatomy of the White Rat
Robert B. Chiasson

Laboratory Anatomy of the Shark
Laurence M. Ashley

Laboratory Anatomy of the Turtle
Laurence M. Ashley

Illustrated by the author

Copyright © 1972, 1979 by
Wm. C. Brown Company Publishers

ISBN 0–697–04629–X

All rights reserved. No part of this publication may be reproduced, stored in a retrieval system, or transmitted in any form or by any means, electronic, mechanical, photocopying, recording, or otherwise, without the prior written permission of the publisher.

Printed in the United States of America
10 9 8 7 6

for Robert K. Enders

# Contents

| CHAPTER | PAGE |
|---|---|
| Preface | viii |
| Introduction | 1 |
| Dissection | 3 |
| Orientation, Planes, and Terminology | 5 |
| 1. Skeleton | 7 |
| 2. Musculature | 16 |
| 3. Digestive and Respiratory Systems | 31 |
| 4. Circulatory System | 35 |
| 5. Urogenital System | 42 |
| 6. Nervous System | 46 |
| 7. Sense Organs | 53 |
| 8. Skin | 55 |

# Preface

In recent years the mink has been gradually replacing the cat as a mammalian dissection specimen in comparative anatomy and vertebrate zoology course laboratories. Increasing demand for preserved cats has been met by a decreasing supply of specimens and vastly increased costs of procuring, housing, and preparing animals in the biological supply houses. Skinned carcasses of the ranch mink are produced by the thousands during the pelting season every year in North America, however. Preserved mink carcasses with double-injected circulatory system can now be bought for less than half the cost of a comparably prepared cat. The mink is a carnivore, like the cat, and many of its structures are similar. The mink differs significantly only in the early fusion of the bones of its skull, the greater complexity of its musculature, the simplicity of its digestive tract, and, of course, in the fact that the skin and parts of the hind feet have been removed during pelting.

The chapter on the skeleton in this manual is based on wild-trapped New England mink. The chapters on the other systems are based on preserved fall-killed ranch mink specimens generously provided by Nasco, of Fort Atkinson, Wisconsin. In previous years mink ranchers usually removed the paws with the skin during pelting. More recently, however, ranchers who sell carcasses to biological supply houses have begun to leave the forepaw and most of the hindpaw on the carcass. Specimens prepared in this fashion can be used for study of the bones of the wrist, forepaw, ankle, and hindpaw, and the musculature of the forearm and the leg. The second edition of the manual includes directions for dissection, descriptions, and illustrations of those structures. In addition, some of the illustrations in the first edition have been replaced by new ones. Selected references to the primary literature of research on the comparative anatomy of mustelid carnivores are included at the end of each chapter. The eye and middle ear of the mink are very small and difficult to dissect. I included a chapter on sense organs, however, should students wish to use supplementary material of a larger mammalian species, such as sheep or cow, for study of the eye, and to attempt a dissection of the ear under a dissecting microscope. A chapter on skin is also included.

A number of users of the first edition suggested modifications which I included in the second edition. I thank in particular Drs. Robert Chiasson (University of Arizona), Emily Oaks (Utah State University), Charles Woods (University of Vermont), and many students at the University of Massachusetts, Amherst.

David Klingener

# Introduction

The mink, *Mustela vison,* is a mammal. The taxonomic position of the animal is as follows:
Phylum — Chordata
  Superclass — Tetrapoda
    Class — Mammalia
      Order — Carnivora
        Family — Mustelidae
          Genus — *Mustela*
            Species — *M. vison*

The Family Mustelidae comprises 25 genera of carnivores, including minks, weasels, ferrets, martens, sables, wolverines, badgers, skunks, otters, and some others. The genus *Mustela* includes the weasels and ferrets in addition to the minks.

Wild minks usually live near water, either streams or marshes, and move long distances in search of food. They eat a wide variety of animals, including fish, frogs, snakes, birds and their eggs, and small mammals, particularly muskrats. They run, swim, and climb well, and seem to be very strong for their size. Males tend to be larger than females. Both males and females have anal scent glands. The musk produced by these glands is considered more offensive than skunk musk by some human connoisseurs.

The mink's lustrous fur coat causes it to be heavily trapped in the wild and to be bred and raised in captivity on mink ranches.

### SUGGESTED READING

Ewer, R. F., 1973. The Carnivores. Cornell University Press: Ithaca, N. Y. 494 pp.

Leonard, A., 1966. Modern Mink Management. Ralston-Purina Company: St. Louis, Missouri. 206 pp.

# Dissection

Dissection is an art and it requires practice. Some students are butchers; their specimens are reduced to vertebral columns and strips of shredded flesh in a matter of days. The opposite fault, timidity, is more common. A structure cannot be understood just by exposing its surface. The attachments of muscles and the full courses of vessels and nerves must be seen. Good dissection lies somewhere between butchery and timidity. It consists mainly of separating and cleaning. Cut structures only when necessary to expose deeper regions, and only after identifying and understanding them fully.

You will need a blunt probe (preferably of the Huber or Bartlett type), a pair of dissection scissors, a pair of forceps, and a scalpel with replaceable blades. A flexible probe for following blood vessels and a bottle opener for working on the skull are useful but not absolutely necessary. Bone cutters and broom straws should be available in the laboratory.

All living things vary. The structures of the mink which you dissect may differ from the descriptions and illustrations in this manual. In any case of difference, the mink is always right.

The drawings in this manual are idealized and in some instances diagrammatic. They are intended to help you work out the structures in the animal. In reviewing before examinations, study the animal, not the illustrations.

# Orientation, Planes, and Terminology

Comparative anatomists use a specialized terminology to describe positions and orientations of structures. In Figure 1 are illustrated the three *planes* on which sections of the animal may be cut. The sagittal plane passing directly through the midline of the animal is the *median sagittal plane.* A sagittal plane parallel to it, but not passing through the midline, is called a *parasagittal plane.*

Most of the terms referring to position and direction are also illustrated in Figure 1. Comparative anatomists tend to use *cranial* and *caudal* rather than *anterior* and *posterior* when referring to the front and hind ends of a quadrupedal animal because the terms *anterior* and *posterior* are used in a different sense in describing structures of the erect human body. The terms *distal* and *proximal* refer to positions of structures on appendages. A structure lying closer to the core, or mass, of the body is said to be *proximal* to a structure lying further out on the appendage. Conversely, the structure lying further out is said to be *distal* to the inner one. Some other terms, such as *superficial* (or *external*) and *deep* (or *internal*) have obvious meanings and are not illustrated.

There are also terms used in referring to regions and structures of the body. Terms such as *head, neck, trunk,* and so forth are obvious. The chest is called

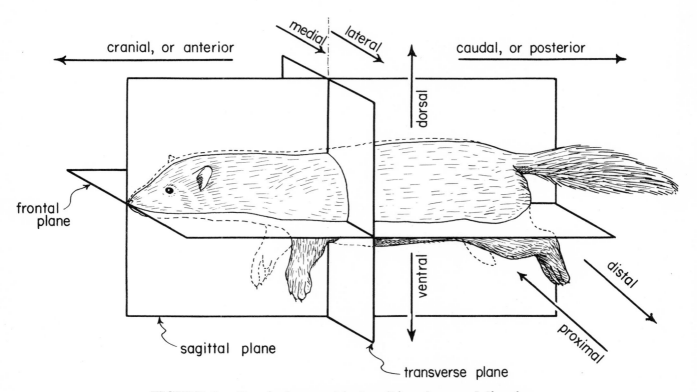

**FIGURE 1. Terminology used in describing planes and directions.**

the *thorax,* and the belly region the *abdomen.* The front limb is the *pectoral limb.* Its segments are the *arm* (between shoulder and elbow), the *forearm* (between elbow and wrist), the *wrist,* and the *forepaw.* The hind limb is called the *pelvic limb,* and its segments are the *thigh* (between hip and knee), the *leg* (between knee and ankle), the *ankle,* and the *hindpaw.*

Movements of the limbs are described as *flexion* (usually closing a joint), *extension* (usually opening a joint), *adduction* (bringing the limb toward the midline), and *abduction* (carrying the limb away from the midline).

The formal names of the structures themselves are based on Latin or Greek roots and usually refer to the appearance, connections, or functions of the structures. By convention comparative antaomists use the names first given to the structures of the human body, which accounts for the seeming inappropriateness of some of the names of the structures of the mink.

# Chapter 1
# Skeleton

Disarticulated skeletons of the mink should be available for study. Handle them carefully. Dried bones, and especially dried teeth, are fragile. When pointing to structures, use a broomstraw instead of a probe or pencil.

The bones and cartilages can be arranged in two series: the *axial skeleton* (skull, mandible, visceral skeleton, vertebral column, ribs, and sternum) and the *appendicular skeleton* (the limbs and their girdles). A small group of bones, the *heterotopic bones,* falls in neither series.

## AXIAL SKELETON

In the mink, as in other mustelids, most individual bones of the skull fuse early during the animal's lifetime. Few if any sutures will be visible in your specimen (Figures 3, 4). Note, however, the large *braincase,* the *orbit* and *temporal fossa,* the strong *mandibular fossa* for articulation with the *condyloid process* of the lower jaw, the large *tympanic bullae,* the bony *secondary palate* which separates the airstream from the mouth cavity, and the other parts and structures illustrated. Pay particular attention to the *foramina.* They transmit ducts, nerves, and blood vessels, and you will encounter them again during your dissection of intact specimens. Each half of the *mandible* is composed of a single bone, the *dentary.* The two halves meet in the midline at the *mental symphysis.* Note that when you look at the mandible in lateral view, a horizontal line passing along the toothrow extends through the condyloid process. This planar relationship of toothrow and condyloid process is characteristic of carnivorous mammals. In herbivores the condyloid process extends far dorsal to the plane of the toothrow.

The visceral skeleton of the mink includes three small bones (*malleus, incus, stapes*) in the middle ear cavity. They transmit vibrations from the eardrum to the inner ear and are not visible in conventionally prepared skulls. The visceral skeleton also includes the *hyoid apparatus* (Figure 5). This develops from ventral elements of the first two visceral arches behind the jaw arch. The base of the tongue and the larynx attach to it (Figure 24). The central skeletal element is the *basihyal,* or *body,* to which are attached two pairs of *horns,* or *cornua.* The *anterior horns* suspend the basihyal from the

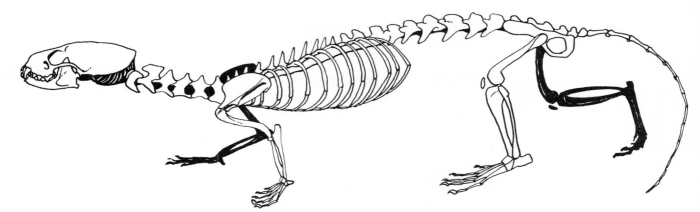

**FIGURE 2.** Articulated skeleton of the mink.

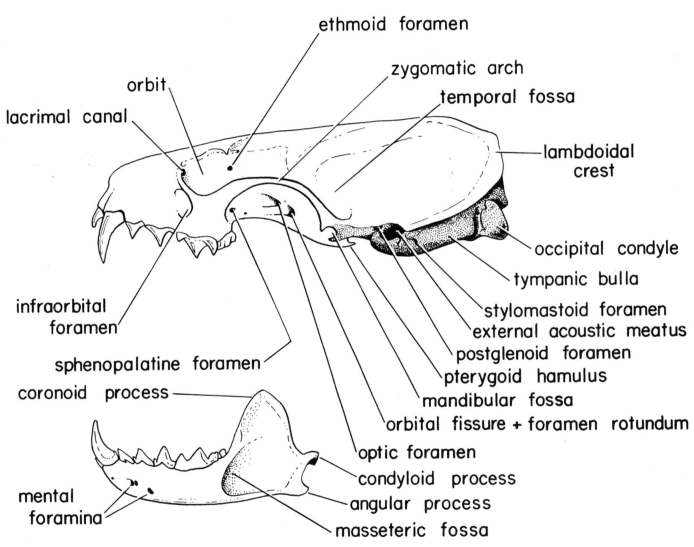

**FIGURE 3.** Skull and mandible, lateral view.

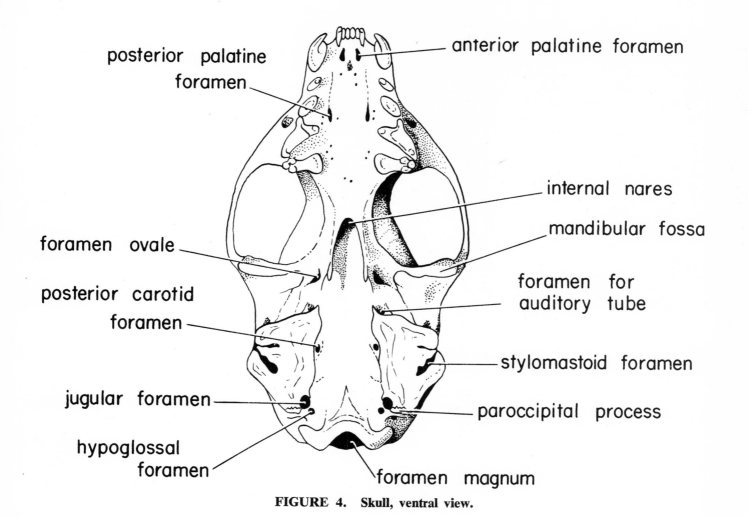

**FIGURE 4.** Skull, ventral view.

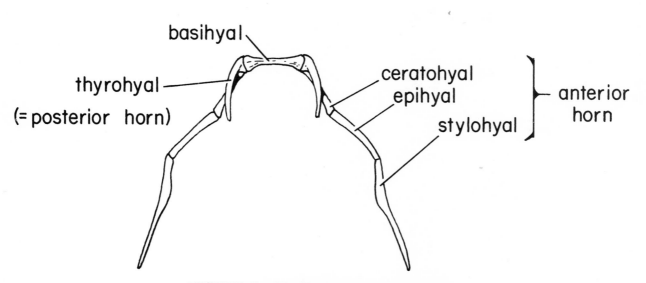

**FIGURE 5.** Hyoid apparatus, ventral view.

skull and consist of three bones each (*ceratohyal, epihyal, stylohyal*) running dorsally from the basihyal to the tympanic bulla and ending short of the stylomastoid foramen. The ceratohyal and epihyal are occasionally fused into a single bone. Each *posterior horn* consists of a single bone, the *thyrohyal*. The cartilages of the larynx are attached to this bone.

The vertebral column of the mink forms a curved, stressed arch from which the trunk is suspended (Figure 2). Morphology of individual vertebrae varies from one region to another within the column, but most vertebrae consist of a *body,* or *centrum,* ventral to a large *vertebral foramen* through which the spinal cord passes (Figure 6). The vertebral foramen is formed by an arch composed of a pair of vertical *pedicles* and a pair of horizontal *laminae*. A *spinous process* protrudes dorsally. Adjacent vertebrae articulate by *cranial* and *caudal articular facets.* The bodies of adjacent vertebrae are separated by *intervertebral discs,* usually lost during preparation of the

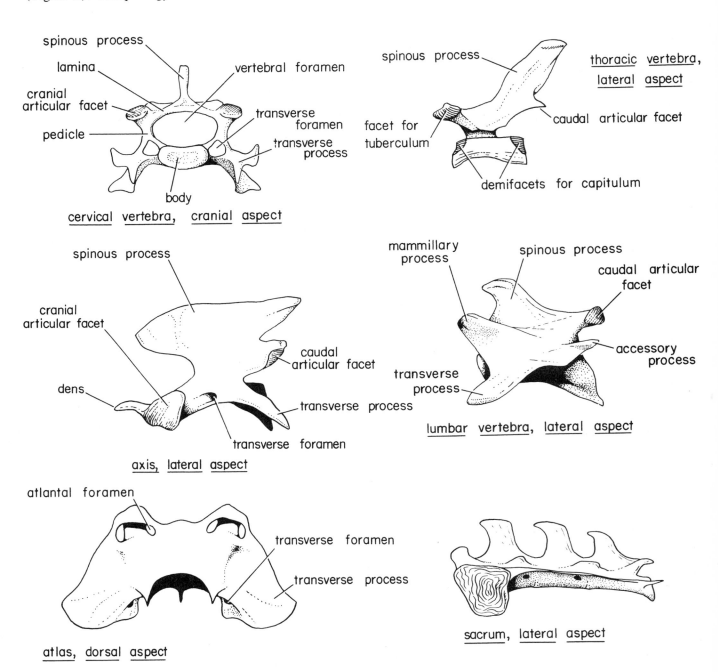

**FIGURE 6. Representative vertebrae.**

skeleton. Most vertebrae have a pair of *transverse processes* extending laterally.

There are seven *cervical vertebrae* in the neck. Each has a pair of *transverse foramina* through which the vertebral arteries and veins run. The first two cervical vertebrae, the *atlas* and *axis,* are highly modified in connection with the movement of the skull on the body. The articulation between the occipital condyles and the atlas permits the head to move in a dorsoventral path only. The movement of the atlas on the *dens* of the axis permits the head to rotate around its long axis. Demonstrate these actions with the atlas, axis, and skull held in articulated position. (In the intact animal a ligament runs transversely across the foramen of the atlas and prevents the dens from contacting the spinal cord.)

There are usually fourteen *thoracic vertebrae*. A pair of ribs attaches to each. The spinous processes of thoracics 1 through 10 incline caudad. The spinous process of thoracic 11 (the *anticlinal vertebra*) is vertical. The spinous processes of the remaining thoracics and all of the lumbars incline craniad. Direction of inclination of spinous processes is affected by muscular stresses acting on the bones as the young animal develops. The first few thoracic vertebrae of the mink may have transverse foramina. This is an unusual situation in mammals, since transverse foramina are usually found only in the first six cervicals. The first ten pairs of ribs (the *true ribs*) articulate ventrally with the sternum by their *costal cartilages*. The costal cartilages of the last four pairs of ribs (the *false ribs*) articulate with the costal cartilages of the true ribs and not directly with the sternum. On each of the anterior ribs is a distinct *tuberculum* and *capitulum* connected by a slender *neck* (Figure 7). The tuberculum articulates with the transverse process of a thoracic vertebra, and the capitulum attaches on the body of the same vertebra and on the body of the next anterior one. Owing to the complex articulation between ribs and vertebrae, movement of the shaft of the rib is both forward and outward during inhalation. On the last few pairs of ribs the tuberculum is absent.

The sternum is formed by eight elements called *sternebrae*. The most cranial element is called the *manubrium,* and the last one is called the *xiphisternum*.

There are six *lumbar vertebrae*. They have large cranially directed transverse processes. *Mammillary* and *accessory processes* are well developed; tendons of the intrinsic muscles of the back (the sacrospinalis) attach to them. Notice the difference in structure and action of the articular facets in the lumbar, thoracic, and cervical series.

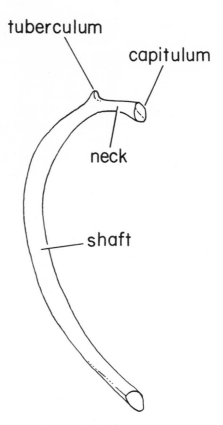

**FIGURE 7.** Rib.

The *sacrum* is formed by the fusion of three vertebrae. Individual spinous processes, intervertebral foramina, and the vestiges of articular processes are visible. There is a large flat process on each side for articulation with the ilium. (The sacrum is so called because Medieval anatomists thought that it was around this bone that each human body would be reconstructed on Resurrection Day—hence, "sacred bone.")

There are eighteen or more *caudal vertebrae*. Vertebral morphology varies greatly from the base of the tail toward the tip, with a strong tendency toward simplification.

## APPENDICULAR SKELETON

The *scapula* or shoulder blade is the major bone of the pectoral girdle (Figure 8). Its flat surfaces (*fossae*) give origin to muscles. The entire medial surface is called the *subscapular fossa*. The lateral surface is divided by the *spine* into *supraspinous* and *infraspinous fossae*. The spine bears a caudally directed *metacromion process* and ends ventrally in the *acromion process*. The scapula articulates with the head of the humerus via the *glenoid fossa*. Near

11

the glenoid fossa is the poorly developed *coracoid process*.

The *clavicle* of the mink is a tiny vestigial bone lying between the clavotrapezius and the clavodeltoid muscles and connected by weak ligaments with the sternum and acromion process. It is usually lost during the preparation of disarticulated skeletons.

The *humerus* is the bone of the arm. It has a rounded *head* proximally for articulation with the glenoid fossa of the scapula and more complex surfaces distally (the *capitulum* and *trochlea*) for articulation with the radius and ulna, respectively. The *greater* and *lesser tubercles, deltoid ridge* and *tuberosity,* and *pectoral ridge* are all developed in areas of major muscle attachment. The *entepicondylar foramen* transmits the brachial artery and the median nerve.

The *radius* and *ulna* are the bones of the forearm. The ulna forms the forearm's major articulation with the humerus, via the *semilunar notch,* and the radius forms the major part of the articulation with the

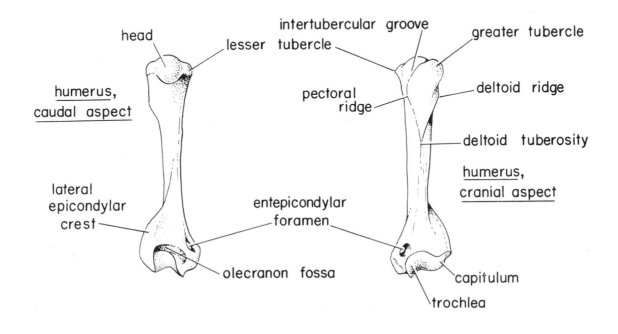

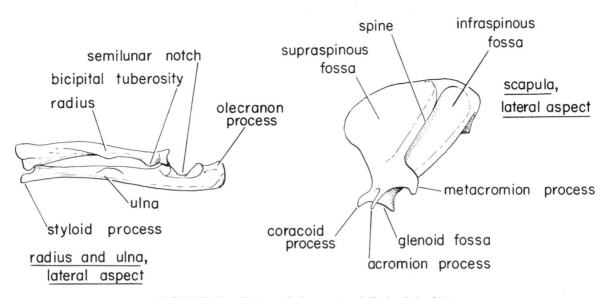

**FIGURE 8. Bones of the pectoral limb (left side).**

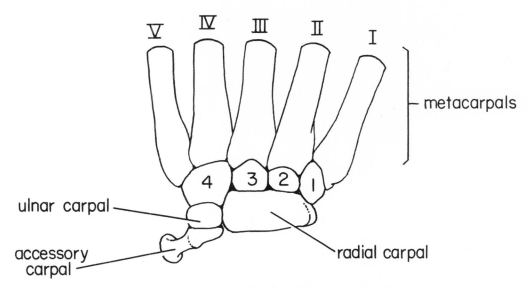

**FIGURE 9.** Bones of the left wrist and forepaw, dorsal view. The cut metacarpals are numbered I through V. The distal carpals are indicated by Arabic numerals (1 through 4).

wrist. The ulna rotates around its own long axis very little, but the round head of the radius is free to rotate. Hence, the actions of pronation and supination of the forepaw are performed mainly by axial rotation of the radius. The *bicipital tuberosity* of the radius and the *olecranon process* of the ulna serve for muscle attachment.

The bones of the wrist and forepaw are best studied in articulated specimens (Figure 9). Terminology used in naming these small bones is complicated, and several different systems are used. The wrist is formed by two rows of carpals. The proximal row includes the *radial carpal* (the fused scaphoid, lunar, and central) and the *ulnar carpal* (the triquetrum). An *accessory carpal* (also called the pisiform) extends ventrally. The distal row includes four bones. The *first carpal* is also called the trapezium, the *second carpal* the trapezoid, the *third carpal* the capitate, and the *fourth carpal* the hamate.

The *digits,* or fingers, are formed by *metacarpals* and *phalanges*. The first digit is formed by one metacarpal and two phalanges. Digits II through V are each formed by a metacarpal and three phalanges.

The *innominate bone* (Figure 10) is composed of three separate bones (*pubis, ischium, ilium*) which fuse early during the animal's lifetime. The left and right pubes and ischia are joined in a *symphysis* at the ventral midline. The *acetabulum* is the socket into which the head of the femur fits.

The *femur* is the bone of the thigh. The conspicuous *head* is joined to the shaft of the bone by the

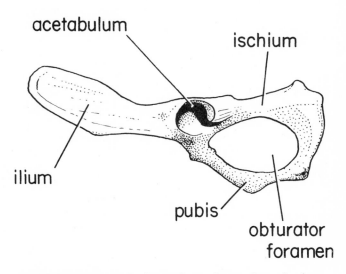

**FIGURE 10.** Left innominate bone, lateral view.

*neck* (Figure 11). In the head is a distinct pit, the *fovea,* in which is attached a strong ligament that holds the head of the femur securely in the acetabulum but permits it to rotate freely. The *greater* and *lesser trochanters* and the *trochanteric fossa* are areas of muscle attachment. Distally the femur bears two *condyles* for articulation with the tibia. A small tubercle above the lateral condyle is used as an aging criterion by wildlife biologists studying mink populations. It is absent in juveniles and first appears at the age of one year. It becomes progressively more distinct as the mink grows older.

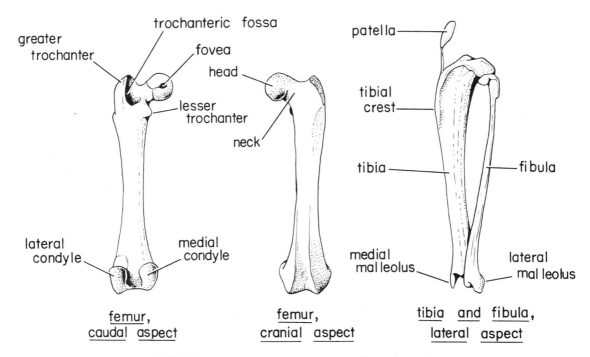

**FIGURE 11.** Bones of the pelvic limb (left side).

The *tibia* and *fibula* are the bones of the leg. The tibia is the large medial bone. The *tibial crest* serves as the point of attachment for the quadriceps femoris group of thigh muscles. Distally the tibia and fibula form *medial* and *lateral malleoli*. These processes resist sideways dislocation of the bones of the ankle.

Bones of the ankle and hindpaw, like those of the wrist and forepaw, are best studied in articulated specimens (Figure 12). The tibia articulates with the *tibial tarsal* (astragalus), and the fibula with the *fibular tarsal* (calcaneus). The tibial tarsal articulates distally with the *central tarsal* (navicular). Four bones form a distal row of tarsals. The first three *tarsals* are also called cuneiformes 1, 2, and 3. The *fourth tarsal* is also called the cuboid. The digits (toes) are formed by *metatarsals* and *phalanges*. The first digit includes a metatarsal and two phalanges. Digits II through V are each formed by a metatarsal and three phalanges.

## HETEROTOPIC BONES

Heterotopic bones ossify in connective tissue, often at points of stress. *Sesamoid bones* may appear in tendons in the regions of joints. The *patella* or kneecap of the mink is a large sesamoid in the tendon of the quadriceps femoris. In mink of both sexes a bone develops in the external genitalia. The *os penis* or *baculum* of the male and the *os clitoridis* of the female will be considered during dissection of the reproductive system.

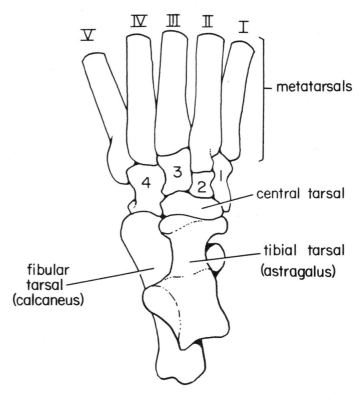

**FIGURE 12.** Bones of the left ankle and hindpaw, dorsal view. The cut metatarsals are numbered I through V. The distal tarsals are indicated by Arabic numerals (1 through 4).

## SUGGESTED READING

Birney, E. C., and E. D. Fleharty, 1968. Comparative success in the application of aging techniques to a population of winter-trapped mink. Southwestern Naturalist, 13:275-282.

Fisher, E. M., 1942. The osteology and myology of the California river otter. Stanford University Press: Stanford, Calif. 66 pp.

Greer, K. R., 1957. Some osteological characters of known-age ranch minks. Jour. Mammalogy, 38:319-330.

Leach, D., 1977. The descriptive and comparative postcranial osteology of marten (*Martes americana* Turton) and fisher (*Martes pennanti* Erxleben): the appendicular skeleton. Canadian Jour. Zool., 55:199-214.

Savage, R. J. G., 1957. The anatomy of *Potamotherium* an Oligocene lutrine. Proc. Zoological Society of London, 129:151-244.

Segall, W., 1943. The auditory region of the arctoid carnivores. Zoological Series of Field Museum of Natural History, 29:33-59.

# Chapter 2
# Musculature

The attachments of a muscle are called the *origin* and *insertion*. Contraction of the muscle draws these two points closer together. The "fixed" attachment, usually closer to the center of the body, is considered the origin. The insertion is located more distally and moves more with respect to the core of the body when the muscle contracts. The muscle's attachment to bone or cartilage may be *fleshy,* by a narrow *tendon,* or by a broad sheet of connective tissue called an *aponeurosis.* The fleshy part of the muscle between the two attachments is called the *belly.*

The skin of the mink has already been removed to serve human thermoregulation and vanity. Dissection of the musculature involves cleaning the fat and fascia off so that fiber directions can be seen, and separating each muscle from adjacent ones. Fingernails and the blunt probe are most useful. When cutting a muscle, first separate it completely from adjacent structures and pass the probe beneath the belly. Then cut downward to the probe with a scalpel from the outside. When cutting a number of muscles that run alongside one another in the same direction (such as the muscles of the thigh), cut across the bellies of different muscles at different levels and reflect (fold back) the muscle halves. Origin and insertion of each muscle can then be rematched after deeper structures are studied, and the superficial parts of the limb reassembled for review.

Dissect the musculature on the left side of the animal only, leaving the right side intact for study of the nervous system. Where two depths of dissection are shown on one animal in the drawings, the deeper is always on the animal's left side.

Refer periodically to the table of muscle origins, insertions, and actions. The trunk muscles have been omitted from the table. The individual actions listed in the table are simplified and very hypothetical. Movements of the limbs and body in any mammal are complex and are produced by the integrated actions of many muscles. Simply stating that an isolated muscle is an "adductor of the thigh" is an oversimplification. Furthermore, the actions of particular muscles can be known for certain only after their activities have been monitored electronically (a process celled *electromyography*) in the living moving animal. Electromyographic studies have never been done on the musculature of the mink.

If one is available, keep a mounted articulated skeleton of the mink in view as you dissect the musculature; it will be easier to visualize muscular attachments and actions.

Like most other mammals, the mink has several muscle sheets associated with the skin. Most of these *dermal muscles* are removed during pelting. Fragments of the *panniculus carnosus* may remain on the back. This muscle twitches the skin over much of the body and is phylogenetically a pectoral muscle. Its fibers converge toward the *axilla* (armpit) and insert with the more orthodox pectoral muscles. Part of the *platysma* may remain on the neck. Its fibers originate in the dorsal midline and sweep downward and forward toward the face. The remaining *superficial facial musculature* on the head is almost always removed with the pelt.

## PECTORAL LIMB

Clean the dermal muscles, fat, and fascia from the neck, back, shoulder, chest, and arm of the left side. Separate the superficial muscles from one another (Figures 13, 17). Cut and reflect the *spinotrapezius, acromiotrapezius,* and *clavotrapezius* close to their origins at the dorsal midline and occiput. The *cleidomastoid* will be found deep to the clavotrapezius posteriorly, sharing its origin on the clavicle. Separate the *pectoralis major* and *minor.* Cut and reflect them separately. Do not damage the underlying blood vessels. The body of the mink is suspended between

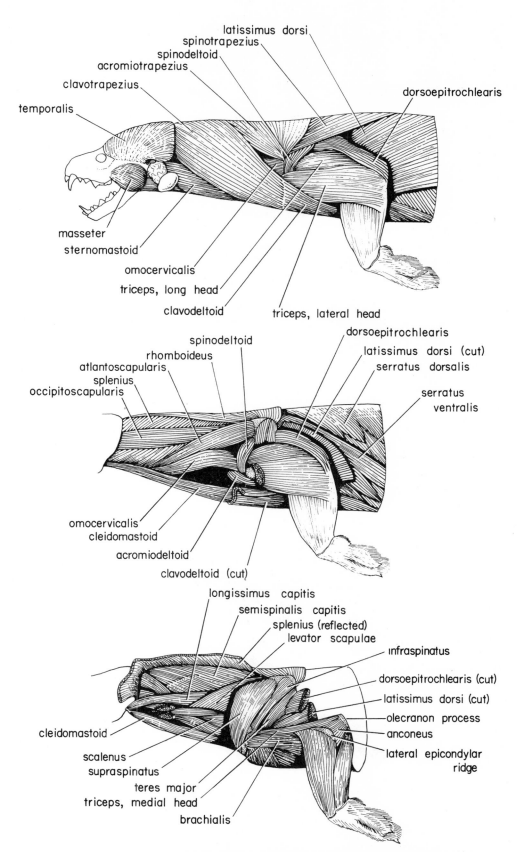

**FIGURE 13.** Muscles of the head, neck, and pectoral limb, lateral view.

the pectoral limbs by several muscular slings. Dorsally, the *rhomboideus, atlantoscapularis, levator scapulae,* and *serratus ventralis* attach to the dorsal edge of the scapula. It is necessary to cut through the rhomboideus close to the dorsal midline and pull the dorsal edge of the scapula away from the body in order to see the serratus ventralis and levator scapulae. The latter two muscles form a continuous sheet, the serratus ventralis originating from ribs and the levator scapulae from transverse processes of cervical vertebrae. The ventral elements of the sling, pectoralis major and minor, attach to the humerus. In primitive mammals the clavicle forms a strong connection between scapula and sternum. In the mink, as in other carnivores and in ungulates, this connection is lost and the clavicle becomes vestigial. Support of the body between the front limbs is then muscular instead of skeletal. The remnants of the mink's clavicle can be found buried in muscle at the junction between the clavotrapezius and the clavodeltoid.

The scapula rotates as the mink runs. Work out for yourself which muscles are responsible for rotation.

The large *latissimus dorsi* and the *teres major* retract the humerus. The humerus is pulled forward by the combined *clavodeltoid and clavotrapezius.* (This combination is frequently referred to as the *brachiocephalic muscle* in carnivores and ungulates.) Other smaller muscles are also involved in these movements of the humerus.

Clean and separate the muscles of the arm (Figures 13, 14). The powerful extensors of the forearm are the heads of the *triceps brachii* and the *dorsoepitrochlearis.* The *anconeus* is not really separable from the medial head of the triceps; it can be viewed as those fibers of the medial triceps that originate on the lateral epicondylar ridge. The major flexors of the forearm are the *biceps brachii* and the *brachialis.*

Musculature of the forearm is complex, and only the larger, more superficial muscles will be described. These muscles are divided into a dorsal *extensor group* and a ventral *flexor group,* each group having its own set of motor nerves. Bellies of the muscles are covered by heavy fascia. Their tendons run distally to insert on bones of the wrist, in the skin of the paw, and on digits of the paw. Separation and identification of the muscles is most easily done by first separating the tendons at the wrist and then passing the blunt probe proximally to separate the muscular bellies in the forearm. As the tendons of both extensor and flexor muscles cross the wrist they are kept in place by heavy transverse bands of connective tissue called *retinacula* (sing., *retinaculum*). The retinacula prevent the tendons from bowing out from the joint during flexion and extension of the paw. Cut the retinacula

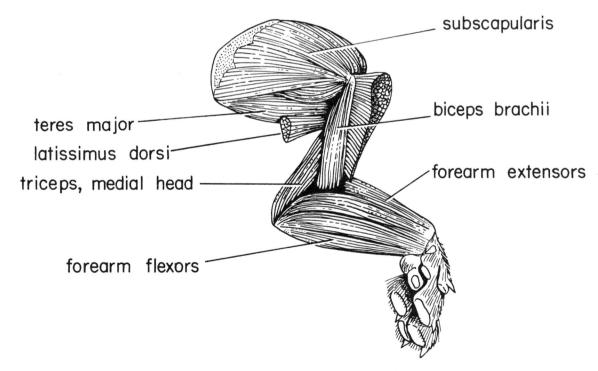

**FIGURE 14.** Muscles of the pectoral limb, medial view.

parallel to the direction of the tendons, being careful to avoid cutting the tendons themselves.

The extensor muscles (Figure 15) originate from the lateral epicondyle, the lateral epicondylar ridge, the humerus proximal to the ridge, and from the dorsal surface of the radius and ulna. Of the extensor muscles, the *brachioradialis* originates most proximally from the humerus. It inserts on the styloid process of the radius. Next to it lie the *extensor carpi radialis longus* (inserts on the metacarpal of digit II) and the *extensor carpi radialis brevis* (inserts on the metacarpal of digit III). Distally these three muscles are crossed by the *abductor pollicis longus*, which inserts on the thumb (digit I). The belly of the *extensor digitorum superficialis* (or communis) lies next to the extensor carpi radialis brevis, and its tendon splits to insert on digits II, III, IV, and V. Next to it lies the belly of the *extensor digitorum lateralis*, which inserts on digits IV and V. The last of the extensor muscles is the *extensor carpi ulnaris*, which inserts on digit V.

One muscle of the flexor group, the *flexor carpi ulnaris*, is visible on the lateral surface of the forelimb. It originates by a number of heads from the humerus and from the olecranon process of the ulna and inserts on the accessory carpal bone. Turning to the ventromedial (flexor) surface of the forearm (Figure 16), identify the *palmaris longus*. The tendon of this slender muscle inserts in the skin of the paw. On each side of the palmaris longus lie the major flexors of the digits, the *flexor digitorum superficialis* and the *flexor digitorum profundus*. Each is composed of a number of heads. The heads of each converge to form single tendons, which split to insert on the phalanges of digits II, III, IV, and V. Next to the tendon of flexor digitorum profundus lies the *flexor carpi radialis*, which inserts on the metacarpals of digits II and III. The *pronator teres* lies next to the flexor carpi radialis and runs between the humerus and the radius. Its action is to pronate the forepaw (turn its ventral side downwards).

A tiny muscle, the *epitrochleoanconeus*, runs between the medial epicondyle of the humerus and the olecranon process of the ulna. It is located just proximal to the pronator teres.

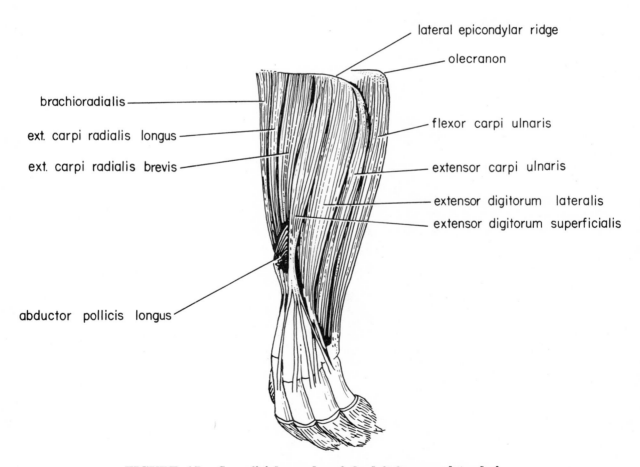

**FIGURE 15. Superficial muscles of the left forearm, lateral view.**

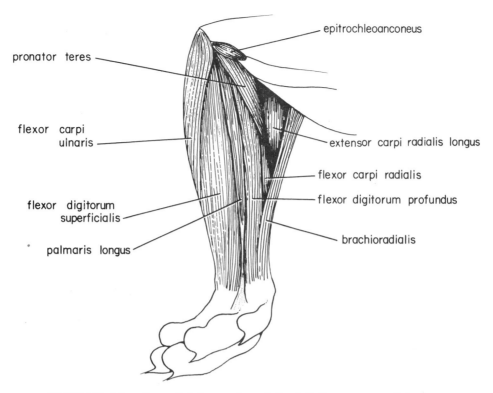

**FIGURE 16.** Superficial muscles of the left forearm, medial view.

The forepaw has intrinsic flexor musculature on its ventral (palmar) surface. There is no intrinsic musculature on the dorsal surface.

## NECK AND HEAD

Cut and reflect the left *sternomastoid*. The slender *sternohyoid*, the larynx, and the trachea will then be exposed. The *thyroid glands* may be found alongside the larynx buried in fat. Pick up the sternohyoid at its insertion on the basihyal. The *sternothyroid*, a slender muscle deep to the sternohyoid, can then be separated. It originates with the sternohyoid on the cartilage of the first rib, and is almost inseparable from it for most of its course in the throat. It inserts on the thyroid cartilage of the larynx, however. The *thyrohyoid* is a short muscle running from the thyroid cartilage to the basihyal. These muscles together position the larynx and the tongue (which is attached to the basihyal). The tongue is pulled strongly back into the mouth when the sternohyoid contracts.

Turning to the head, identify the powerful *temporalis* and *masseter muscles*. Note that the pull of the temporalis on the mandible is upward and backward, and the pull of the masseter is upward and forward. The temporalis is almost always much larger than the masseter in carnivores. The deeper *pterygoid muscles* also affect the position of the mandible, but they cannot be seen until the deeper structures of the head have been dissected. The *digastric muscle* opens the mouth. It has a *tendinous inscription* across it separating it into two parts. The muscle is of dual embryonic origin, the anterior belly coming from the mandibular arch and the posterior belly from the hyoid arch.

Cut across the left digastric muscle and reflect both bellies. The *mylohyoid*, which forms the muscular floor of the mouth, is now entirely visible (Figures 18, 24). Cut the mylohyoid down the midline and reflect the left half. Separate the underlying muscles without damaging the salivary ducts, nerves, veins, and arteries. The slender muscle close to the midline immediately deep to the mylohyoid is the *geniohyoid*. It runs from the mental symphysis to the *basihyal*. It protrudes the tongue; its action is opposite to that of the sternohyoid. A number of *extrinsic tongue muscles* run from skeletal elements into the tongue itself: the *genioglossus* from the mental symphysis, the *hyoglossus* from the basihyal, and the *stylogossus* from the anterior horn of the hyoid apparatus and the surface of the tympanic bulla. The tongue also has a set of *intrinsic muscles* confined to the body of the tongue itself. These control the shape of the tongue during feeding. They need not be dissected.

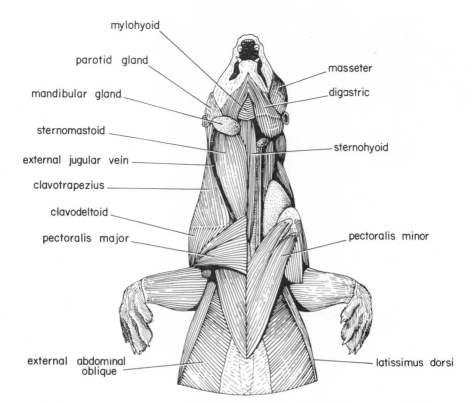

**FIGURE 17.** Muscles of the head, neck, and pectoral limb, ventral view.

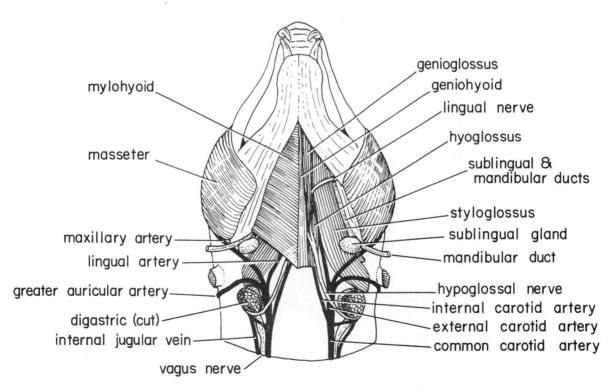

**FIGURE 18.** Muscles, nerves, blood vessels, and ducts of the head, ventral view.

## TRUNK

The lateral abdominal muscles of mammals consist of three superimposed sheets whose fibers run in different directions. The outermost is the *external abdominal oblique,* whose fibers start dorsally and run posteroventrad. Slit and peel off a small section of the external abdominal oblique on the middle of the left flank, about halfway up the body. Under it you should see the *internal abdominal oblique.* Its fibers originate dorsally and run anteroventrad. Slit and reflect a section of this sheet to find the *transversus abdominis.* Its fibers run almost directly dorsoventrally. Inside this sheet is the *peritoneum* and the abdominal cavity.

Ventrally the abdominal muscle sheets become fascial as they approach the midline. The fascial sheets form a complex *sheath* for the *rectus abdominis.* The rectus abdominis runs from the innominate bone to the first rib and is the major support of the abdominal viscera. Some anatomists also view the rectus and its sheath as a bowstring, with the vertebral column, innominate bone, and anterior ribs as the bow. The tendency of the "bow" to straighten, caused by the vertebral ligaments and muscles, is opposed by the rectus and sheath, hence the vertebral column retains its arch permanently.

In the thorax the abdominal sheets are in effect interrupted by ribs and are represented as the *external* and *internal intercostals* and the *transversus thoracis.* These muscles can be seen better by dissection from inside the chest cavity and can be studied after the digestive, respiratory, and circulatory systems have been dissected.

Find the *serratus dorsalis* and the *scalenus.* These run from the vertebral column to the ribs. They are thought to be respiratory muscles, pulling the rib cage forward and outward during inspiration. The sternomastoid muscle may possibly also act as a respiratory muscle, pulling the sternum forward if the head is fixed in position by other muscles.

The intrinsic musculature of the back is the most complex set of muscles in the mammalian body (Figure 19). It consists of numerous bundles of fibers attaching on the ilium, processes of vertebrae,

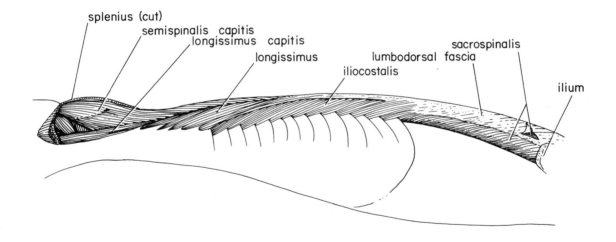

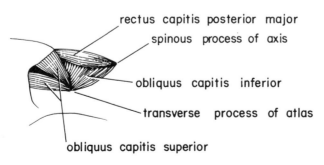

**FIGURE 19. Intrinsic muscles of the back, lateral views.**
**Above:** musculature after removal of the splenius.
**Below:** the suboccipital muscles after removal of the semispinalis capitis.

ribs, and the occipital region of the skull. These muscles bend the vertebral column from side to side and also dorsally. Dorsoventral flexion and extension of the column are an important component of locomotion in carnivores, adding significantly to stride length. The muscle mass is poorly differentiated in the posterior part of the spinal column and reaches its greatest differentiation anteriorly in the cervical region. In the lumbar region the intrinsic muscles of the back can be considered a single mass, the *sacrospinalis*. Fibers take origin from the ilium, from the mammillary, accessory, transverse, and spinous processes of lumbar vertebrae, and from the thick *lumbodorsal fascia*. Part of the sacrospinalis extends out on the dorsum of the tail as the *caudal extensor muscles*. In the thoracic region the sacrospinalis divides into three longitudinal columns. The lateralmost is the *iliocostalis*, whose fibers insert on ribs. The column lying medial to it is the *longissimus*, which inserts on ribs and on the transverse processes of cervical vertebrae. A separate division in the cervical region, *longissimus capitis*, extends forward to insert on the skull. The most medial column is the *transversospinalis*. In the thorax it consists of many indistinct small bundles connecting transverse processes of vertebrae with transverse processes and spinous processes of other vertebrae. In the cervical region the transversopinalis mass differentiates into the prominent *semispinalis capitis*, which attaches to the skull and lies deep to the *splenius*. A highly specialized muscle group, the *suboccipital muscles*, controls movement of the skull, atlas, and axis. The *rectus capitis posterior major* lies deep to the semispinalis capitis and runs between the spinous process of the axis and the skull. Beneath it can be found the *rectus capitis posterior minor* (not shown in Figure 19) connecting the atlas and the skull. The *obliquus capitis inferior* runs between the spinous process of the axis and the transverse process of the atlas; the *obliquus capitis superior* is a two-part muscle connecting the transverse process of the atlas and the skull.

There are also muscles on the ventral surface of the vertebral column that bend it ventrally (*longus colli* and *longus capitis* in the neck; *psoas minor* in the lumbar region). They can be seen after the viscera have been dissected later.

## PELVIC LIMB

Remove the fat and fascia from the hind limb and posterior part of the body. If your specimen is a male, leave the fat in the inguinal region; the testicles are buried in it. Separate and identify the superficial muscles (Figures 20, 21), cut and reflect the *biceps femoris* and *sartorius* close to their insertions, and cut and reflect the *gluteus maximus* and *tensor fascia lata*. On the medial surface of the thigh, cut and reflect the *gracilis*. Identify all the muscles now exposed.

The femur is pulled backward (extended) by the *gluteus medius* and *minimus*, by the powerful *hamstring muscles* (*biceps femoris, semitendinosus, semimembranosus,* and *presemimembranosus*) and to some extent by the *adductors*. The hamstring muscles, except for the presemimembranosus are "two-joint" muscles since they cross both the hip and knee joints. Normally, when they contract they cause extension of the femur and flexion (backward movement) of the tibia. The femur is flexed (pulled forward) by the *sartorius* and *rectus femoris*. These are also two-joint muscles and extend the tibia in addition to their action on the femur. The tibia is also extended by the *vastus lateralis, medialis,* and *intermedius*. The vasti and the rectus femoris are collectively termed the *quadriceps femoris*.

The femur is abducted by the *tensor fascia lata, gluteus maximus,* and *femorococcygeus*, and adducted by the *adductors*.

Deeper dissection will reveal a complex group of muscles on the dorsal edge of the ischium. The *obturator internus* originates on the medial surface of the membrane covering the obturator foramen and passes dorsolaterally over the ischium. The *gemellus superior* and *inferior* originate on the ischium anterior to and posterior to the point where the internal obturator crosses the bone and insert together with the internal obturator in the trochanteric fossa. Together they rotate the femur.

Morphologically the hind limb musculature of mustelids is rather primitive by comparison with other carnivores. In the cat, for example, the adductor magnus and brevis are fused into a single muscle, the presemimembranosus and semimembranosus are fused, and the semimembranosus has begun to fuse with the adductor group. The tenuissimus of the cat is also considerably reduced.

Many of the names applied to pelvic and thigh muscles seem to be inappropriate. The mink's gluteus maximus is smaller than its gluteus medius, and its adductor brevis is longer than its adductor longus. The muscles were named first in the human where they look rather different owing to the reorganization of hind limb musculature when man became bipedal.

Dissection of the muscles of the leg resembles the approach to dissection of the forearm. Remove the heavy fascia and fat first. Then identify the tendons as they pass over the ankle. When the tendons are separated and identified, pass your probe proximally to separate the bellies of the muscles from one another. Like the muscles of the forearm, the muscles

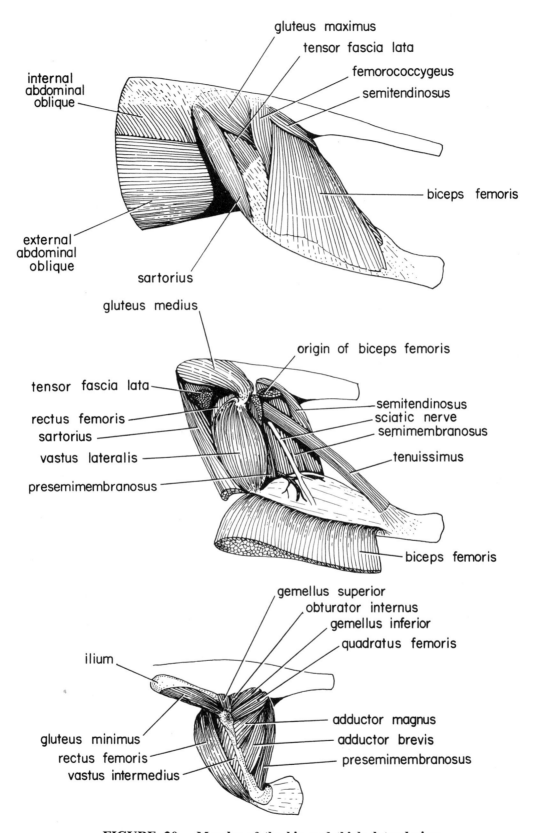

**FIGURE 20.** Muscles of the hip and thigh, lateral view.

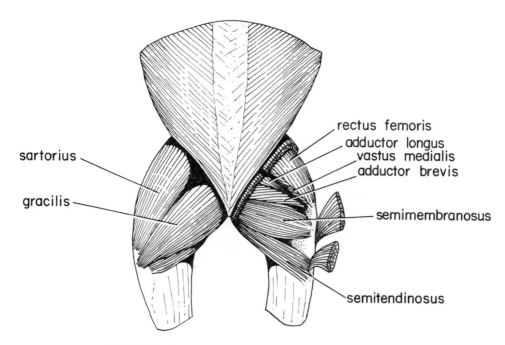

**FIGURE 21.** Muscles of the thigh, ventral view.

of the leg are divided into a dorsal *extensor mass,* which extends the digits of the hindpaw (this action also being called dorsiflexion), and a ventral *flexor mass,* which flexes the digits. The tibia is exposed craniomedially, establishing a convenient boundary between extensor and flexor groups. Note that the bulk of the flexor muscles is collectively much greater than that of the extensors. The flexors are involved in the power stroke during locomotion, and the extensors in the recovery stroke.

The largest flexor muscle is the *gastrocnemius* (Figures 22, 23). The *medial head* originates from the medial epicondyle of the femur, and the *lateral head* from the lateral epicondyle. The two heads join to form the heavy *calcaneal tendon* (tendon of Achilles), which inserts on the tip of the calcaneus. Deep to the lateral head is the *flexor digitorum superficialis* (plantaris), which runs distally along with the gastrocnemius. Its tendon passes medially around the calcaneal tendon and runs behind the tip of the calcaneus to the sole of the hindpaw, where it splits to insert on digits II through V. Cut and reflect the gastrocnemius and the flexor digitorum superficialis. The popliteus is a fan-shaped muscle originating on the lateral epicondyle of the femur. It inserts on the tibia and rotates the leg. Distal to its insertion lies the *flexor digitorum profundus.* This muscle has two parts: the *flexor digitorum longus* forms a tendon which passes behind the medial malleolus of the tibia and inserts on digits II through V, and the *flexor hallucis longus* forms a thinner tendon that also passes behind the medial malleolus but inserts on digit I.

The largest extensor muscle is the *tibialis cranialis* (tibialis anterior). It originates on the cranial surface of the tibia, and its large tendon passes under a heavy retinaculum to insert on metacarpal I. The tendon of the *extensor digitorum longus* passes beneath the same retinaculum and splits to insert on the phalanges of digits II through V. Deep to these is the slender *extensor hallucis longus,* which originates on the fibula and distally forms a slender tendon that runs out along the dorsum of digit I.

Between these extensors and the flexor mass lies the *peroneal group* of three extensors. Most superficial is the *fibularis longus* (peroneus longus), whose tendon passes in a groove over the lateral malleolus of the fibula. The tendon then passes onto the sole of the hindpaw to insert on digit I. The tendons of the other two muscles, *fibularis brevis* (peroneus brevis) and *extensor digitorum lateralis* (peroneus tertius), pass behind the lateral malleolus in a common sheath. They both insert on digit V, the fibularis brevis by a stout tendon onto the base of the metatarsal, and the extensor digitorum lateralis by a slender tendon that runs out along the dorsum of the digit.

Like the forepaw, the hindpaw has intrinsic flexor muscles. Unlike the forepaw, the hindpaw also has some intrinsic extensors.

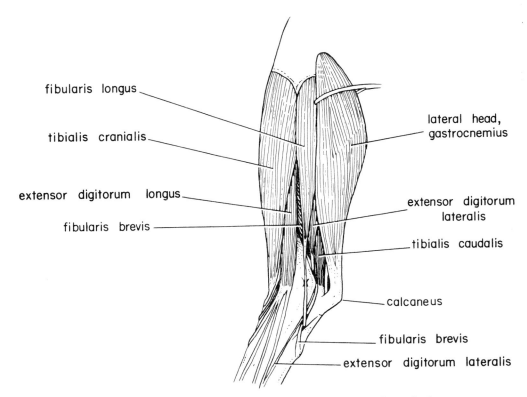

**FIGURE 22.** Superficial muscles of the left leg, lateral view.

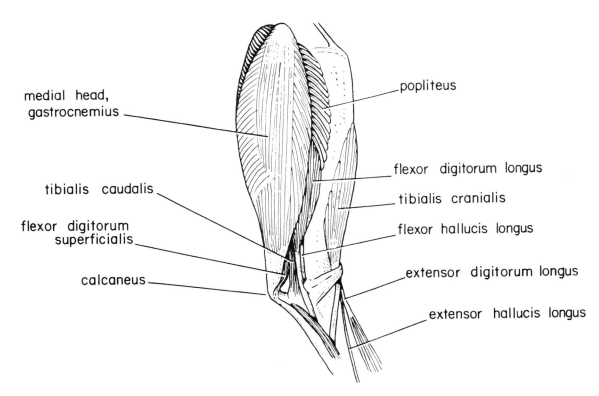

**FIGURE 23.** Superficial muscles of the left leg, medial view.

## PECTORAL LIMB

| NAME | ORIGIN | INSERTION | ACTION |
|---|---|---|---|
| Clavotrapezius | Occiput and dorsal midline of neck | Clavicle (and clavodeltoid) | Pulls humerus forward |
| Acromiotrapezius | Dorsal midline of neck | Acromion process | Pulls scapula dorsad |
| Spinotrapezius | Dorsal midline of thorax | Scapular spine | Rotates scapula backward |
| Pectoralis major | Sternum | Pectoral ridge and part of humerus distal to it | Pulls humerus forward and adducts it |
| Pectoralis minor | Sternum | Pectoral ridge of humerus | Adducts humerus |
| Latissimus dorsi | Dorsal fascia of thorax; also some posterior ribs | Medial surface of humerus | Pulls humerus backward |
| Teres major | Posteroventral border of scapula | Medial surface of humerus | Pulls humerus backward |
| Dorsoepitrochlearis | Surface of latissimus dorsi and teres major | Olecranon process of ulna | Extends forearm |
| Spinodeltoid | Scapular spine | Deltoid ridge of humerus | Pulls humerus forward and abducts it |
| Acromiodeltoid | Acromion process | Deltoid ridge and tuberosity | Pulls humerus forward |
| Clavodeltoid | Clavicle (and clavotrapezius) | Humerus distal to deltoid tuberosity | Pulls humerus forward |
| Rhomboideus | Dorsal midline of thorax and neck | Dorsal edge of scapula | Pulls scapula upward and inward |
| Occipitoscapularis (= Rhomboideus capitis) | Occiput | Dorsal edge of scapula | Pulls scapula forward |
| Atlantoscapularis | Transverse process of atlas | Root of scapular spine | Rotates scapula forward |
| Omocervicalis (= Levator scapulae ventralis) | Transverse process of atlas | Metacromion process | Pulls scapula forward |
| Supraspinatus | Supraspinous fossa | Greater tubercle of humerus | Rotates arm |
| Infraspinatus (includes teres minor) | Infraspinous fossa | Greater tubercle of humerus | Abducts and rotates arm |
| Subscapularis | Subscapular fossa | Lesser tubercle of humerus | Adducts arm |
| Serratus ventralis | Ribs | Dorsal border of scapula | Depresses scapula; supports trunk |
| Levator scapulae | Transverse processes of last 5 cervical vertebrae | Dorsal border of scapula | Depresses scapula; supports trunk |
| Biceps brachii | Tubercle on scapula | Bicipital tuberosity of radius | Flexes forearm |
| Brachialis | Lateral surface of humerus | Ulna | Flexes forearm |
| Triceps brachii: | | | |
| long head | Scapula | Olecranon process | Extends forearm |
| lateral head | Lateral surface of humerus | Olecranon process | Extends forearm |
| medial head | Posterior surface of humerus | Olecranon process | Extends forearm |
| Anconeus | Lateral epicondylar ridge | Olecranon process | Extends forearm |

## HEAD AND NECK

| NAME | ORIGIN | INSERTION | ACTION |
|---|---|---|---|
| Sternomastoid | Sternum | Occiput | Turns head |
| Cleidomastoid | Clavicle | Occiput | Turns head |
| Sternohyoid | Cartilage of first rib | Basihyal | Pulls tongue backward |
| Sternothyroid | Cartilage of first rib | Thyroid cartilage of larynx | Pulls larynx backward |
| Thyrohyoid | Basihyal | Thyroid cartilage of larynx | Pulls larynx forward |
| Mylohyoid | Medial surface of mandible | Basihyal; also meets its fellow in midline | Elevates floor of mouth |
| Digastric | Paroccipital process and mastoid | Medial surface of mandible | Depresses mandible |
| Geniohyoid | Mental symphysis | Basihyal | Pulls tongue forward |
| Genioglossus | Mental symphysis | Tongue | Positions tongue |
| Styloglossus | Anterior horn of hyoid apparatus, and tympanic bulla | Tongue | Positions tongue |
| Hyoglossus | Basihyal | Tongue | Positions tongue |
| Temporalis | Temporal fossa | Coronoid process of mandible | Elevates mandible |
| Masseter | Zygomatic arch | Masseteric fossa of mandible | Elevates mandible |
| Pterygoideus externus | Skull behind last molar | Medial surface of mandible | Elevates mandible |
| Pterygoideus internus | Pterygoid fossa and hamulus | Medial surface of mandible | Elevates mandible |
| Scalenus | Transverse processes of cervical vertebrae | Several anterior ribs | Pulls rib cage forward |
| Longus colli | Cervical and anterior thoracic vertebrae | Cervical vertebrae as far forward as atlas | Bends neck downward |
| Longus capitis | Ventral surfaces of second through sixth cervical vertebrae | Ventral surface of occipital bone | Bends head downward |

## PELVIC LIMB

| NAME | ORIGIN | INSERTION | ACTION |
|---|---|---|---|
| Gluteus maximus | Fascia over sacrum | Lateral surface of femur | Abducts thigh |
| Tensor fascia lata | Ilium | Fascia lata of thigh | Abducts thigh |
| Femorococcygeus (=Caudofemoralis) | Fascia over root of tail | Lateral surface of femur | Abducts thigh |
| Gluteus medius | Ilium and fascia over sacrum | Greater trochanter | Extends thigh |
| Gluteus minimus | Ventral part of ilium | Greater trochanter | Extends femur |
| Biceps femoris | Ischium | Lateral surface of knee and leg | Extends thigh and flexes leg |
| Tenuissimus | Fascia over root of tail | Lateral surface of leg | Flexes leg |
| Semitendinosus | Ischium | Medial surface of leg | Extends thigh and flexes leg |
| Semimembranosus | Ischium | Medial surface of knee and leg | Extends thigh and flexes leg |
| Presemimembranosus | Ischium | Femur above condyles | Extends thigh |
| Gracilis | Pubis and ischium | Medial surface of knee and leg | Adducts thigh |
| Adductor longus | Pubis | Femur | Adducts thigh |
| Adductor brevis | Pubis | Femur | Adducts thigh |
| Adductor magnus | Ischium | Femur | Adducts thigh |
| Obturator externus | Lateral surface of obturator foramen | Trochanteric fossa | Rotates thigh |
| Sartorius | Ilium | Knee | Flexes thigh, extends leg |
| Rectus femoris | Ilium | Tibial crest | Flexes thigh, extends leg |
| Vastus lateralis | Anterior surface of femur | Tibial crest | Extends leg |
| Vastus medialis | Medial surface of femur | Tibial crest | Extends leg |
| Vastus intermedius | Anterior surface of femur | Tibial crest | Extends leg |
| Pyriformis | Transverse processes of second and third sacrals | Greater trochanter | Extends thigh |
| Gemellus superior | Dorsal border of ischium | Trochanteric fossa | Rotates thigh |
| Gemellus inferior | Dorsal border of ischium | Trochanteric fossa | Rotates thigh |
| Obturator internus | Inner surface of obturator foramen | Trochanteric fossa | Rotates thigh |
| Quadratus femoris | Ischial tuberosity | Greater trochanter | Rotates thigh |
| Psoas minor | Ventral surfaces of bodies of last few thoracic and first few lumbar vertebrae | Ilium | Flexes spine |
| Iliopsoas | Bodies of last few thoracic and all lumbar vertebrae | Lesser trochanter | Flexes thigh |

## SUGGESTED READING

Bisaillon, A., 1976. La Musculature du Membre pelvien du Putois d' Amerique (*Mustela nigripes* Audubon et Bachman). Anatomischer Anzeiger, 139:486-504.

Fisher E. M., 1942. The osteology and myology of the California river otter. Stanford University Press: Stanford, Calif. 66 pp.

Hall, E. R., 1926. The muscular anatomy of three mustelid mammals, *Mephitis, Spilogale,* and *Martes.* Univ. California Publ. in Zoology, 30:7-38.

Windle, B. C. A., and F. G. Parsons, 1897. On the myology of the terrestrial Carnivora—part I. Muscles of the head, neck, and fore-limb. Proc. Zoological Society of London, 1897:370-409.

———. *On the myology of the terrestrial Carnivora—part* II. Proc. Zoological Society of London, 1898:152-186.

# Chapter 3
# Digestive and Respiratory Systems

## SALIVARY GLANDS

There are five pairs of salivary glands in the head (Figures 17 and 18). The *parotid gland* is a flat, indistinct mass lying just below the ear. Its duct runs forward on the surface of the masseter muscle and enters the mouth near the last upper premolar. The *mandibular gland* is a round, smooth-surfaced body lying ventral to the parotid gland. Its duct runs forward within the musculature of the floor of the mouth and opens into the mouth cavity near the base of the tongue. The *sublingual gland* lies deeper; deep to the digastric muscle, caudal to mylohyoideus, and lateral to styloglossus. Its duct parallels the duct of the mandibular gland and opens at the same spot. The other salivary glands, the *buccal* and *zygomatic glands*, are very small and are located in the cheek and orbit, respectively. They may be found later during deeper dissection of the head.

## MOUTH CAVITY

Using the bone cutters, cut the left mandible just behind the mental symphysis. Reflect the lower jaw, exposing the mouth. Locate the *oral cavity* (mouth cavity proper) and the *vestibule* (the space between lips and teeth). Note the ridges on the roof of the mouth. Expose the tongue (Figure 24). It is attached to the floor of the mouth by a fold of connective tissue, the *frenulum*. The tongue's upper surface is covered by numerous *papillae*. Most of these are spine-like *filiform papillae*. A few are mushroom-shaped *fungiform papillae*. At the base of the tongue are two pairs of large *vallate papillae* arranged in a V. Each vallate papilla is surrounded by a circular groove.

## TEETH

These are best studied in the cleaned skeletal material. In each side of its head the mink has three *incisors* in upper and lower jaw, a single *canine* above and below, and three *premolars*. The last upper premolar is a large tooth with a flat cutting blade. It occludes with the blade of the large *first molar* of the lower jaw. These particular teeth are called *carnassials*, and they are used by carnivores to cut tough meat, tendons, cartilage, and bone. The carnassials are in the posterior part of the jaw, close to the mandibular joint and muscles. Behind the large last premolar in the upper jaw is a single molar. In the lower jaw, the large first molar is followed by a very small second molar. The dentition of the mink is specialized for feeding on flesh; for puncturing, tearing, and cutting food material. Very little grinding is done, and the molar series of teeth is reduced by comparison with omnivorous and herbivorous mammals.

## PHARYNX

The caudal boundary of the oral cavity is formed by the *pillars of the fauces,* a pair of connective tissue and muscular folds that arch from the base of the tongue to the palate. The region between the pillars of the fauces and the esophagus is the *pharynx*. In the pharynx the air passage moves from its dorsal position in the nasal cavity to its ventral position in the trachea, thereby crossing the food passage. The pharynx is divisible into three regions. The *nasopharynx* lies above the soft palate. It transmits air. In its walls are the slit-like *openings of the auditory tubes* (or *Eustachian tubes*) which lead to the right and left middle ear cavities. They are very small openings in the mink and are difficult to see without a dissecting microscope. The *oropharynx* lies between the pillars of the fauces and the basihyal. In its walls are the flap-like *palatine tonsils,* lying just craniad of the border of the soft palate. The *laryngopharynx* is the pharyngeal region lying just above the larynx.

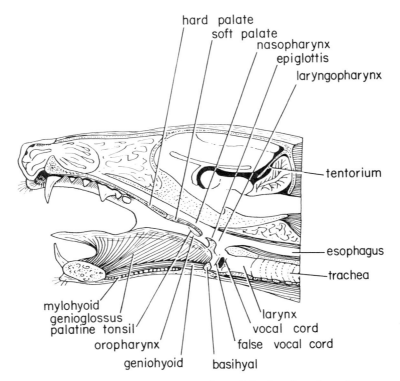

**FIGURE 24.** Sagittal section of the head and neck.

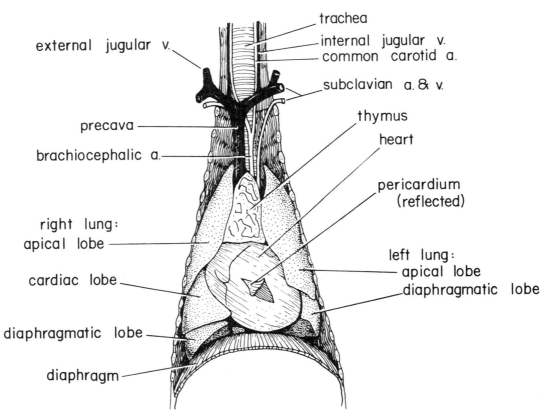

**FIGURE 25.** Viscera of the thorax.

## NECK

Note that the *esophagus* is normally collapsed. The *trachea*, on the other hand, is held open by a series of cartilaginous rings in its wall. Cut the *larynx* midventrally and spread the two halves apart. The larynx is formed by the large *thyroid cartilages,* the smaller *cricoid cartilages,* and the minute *arytenoid cartilages* on its dorsal surface. On its inner wall can be seen the two pairs of *vocal cords*. The caudal pair are the true vocal cords and are thought to produce sound by vibration as the air passes over them. The cranial pair, the false vocal cords, are not thought to be involved in sound production. The laryngeal cartilages are moved upon one another by the *intrinsic muscles of the larynx*. As the cartilages move, tension of the vocal cord is increased and decreased, and the pitch of the cry varies.

## THORACIC CAVITY

Open the *thoracic cavity* (or chest cavity) by cutting through the muscles and rib cartilages on the left side of and parallel to the sternum. Keep the scissors pointed ventrally (toward yourself) as much as possible to avoid damaging structures in the cavity. Pull the walls of the cavity laterad, breaking the ribs. The *thymus gland* is a mass of dark brown tissue embedded in the fat cranial to the heart (Figure 25). Carefully remove the thymus and fat from around the major organs. Use a probe and forceps instead of a scalpel; the blood vessels and nerves must not be damaged. The heart lies in the *pericardial cavity,* delimited by the tough *pericardium*. The lungs lie in the *pleural cavities,* the other subdivisions of the thoracic cavity. The right lung has three major lobes, the *apical, cardiac,* and *diaphragmatic,* and a fourth smaller lobe, the *intermediate,* more dorsal in position and associated with the postcava. The left lung has two lobes, the *apical* and *diaphragmatic*. Follow the trachea and esophagus as they enter the thorax. Dorsal to the heart, the trachea divides into left and right *bronchi,* which carry air to and from the lungs. Defer dissection of this region until after removal of the heart during study of the circulatory system. The esophagus continues dorsal to the heart and penetrates the muscular *diaphragm* to enter the *abdominal cavity*. The periodic contractions of the diaphragm, together with the forward and outward movement of the ribs, increase the volume of the pleural cavities and cause inspiration of air into the lungs.

## ABDOMINAL CAVITY

Open the abdominal cavity by making a single incision through the ventral body wall from the end of the sternum to the pubis. Cut the body wall also along the edges of the rib cage and reflect the muscle sheets laterally to expose the viscera. Anteriorly, the dark lobes of the *liver* should be visible (Figure 26). The

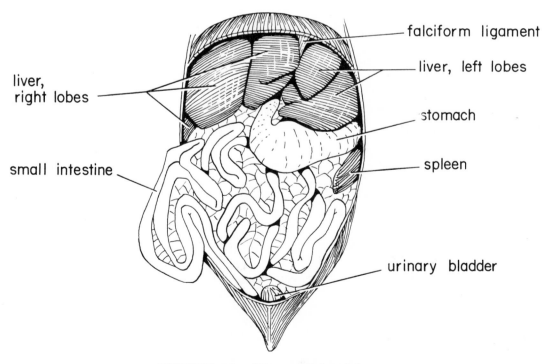

**FIGURE 26.** Viscera of the abdomen.

mesentery between the liver and the diaphragm is the *falciform ligament,* which divides the liver into right and left sides. The lobe of the right side of the liver closest to the midline (the right median lobe) contains the dark green *gall bladder.*

Identify the stomach (Figure 27). The stomach is attached to the liver and part of the small intestine by a mesentery called the *lesser omentum.* Attached to the greater curvature of the stomach is the *greater omentum,* an extensive sheet of mesentery laden with fat. It extends caudally and covers most of the remaining abdominal viscera. Cut the greater omentum near its attachment to the stomach and remove it. Try to keep all the other mesenteries intact. Identify the regions and parts of the stomach and cut it open to expose its inner surface. Note the *rugae,* the large longitudinal ridges. Size of the stomach in the mink, as in other carnivores, depends on how recently and how well the individual ate. If the stomach in your animal is full of food, it may be enormous. The stomach is closed by contraction of the *pyloric sphincter.* When the sphincter relaxes, food is permitted to pass into the small intestine.

The *spleen* is a greenish-brown organ lying in a mesentery on the left side of the stomach. It is part of the circulatory system.

Identify the *small intestine,* which begins at the pyloric sphincter. In the mesentery of the first part of the small intestine lies the *right limb of the pancreas.* It is pinkish and rather loose in structure. The *left limb of the pancreas* lies near the stomach and extends to the spleen. The products of the pancreas (digestive enzymes) and of the liver (bile salts) are carried into the small intestine by a common duct system. Find the large *cystic duct* from the gall bladder and several *hepatic ducts* from the liver. These join to form the *common bile duct.* Bile passes from the liver to the gall bladder, where it is stored and concentrated. Eventually it is emptied into the small intestine. The common bile duct enters the small intestine near the pylorus, and its point of entry may be marked internally by a small *papilla.* The two *pancreatic ducts,* one from each limb, join the common bile duct just before it enters the small intestine. Occasionally one of the pancreatic ducts will have a separate entry to the intestine.

The small intestine is divided into three segments: the *duodenum,* which begins at the pyloric sphincter, the *jejunum,* and the *ileum.* They are best distinguished by histological study of their walls. The ileum opens into the *large intestine,* or *colon.* There is no *cecum,* or pouch, developed at this point in the gut of the mink. The colon is not divisible into *ascending, transverse,* and *descending* segments as in many other mammals. It is instead a short descending tube that ends in the *rectum.* Defer dissection of the rectum until you study the urogenital system. If the colon is empty, its diameter may be no larger than that of the small intestine, and locating the *ileocolic junction* (the boundary between the large and small intestine) may be difficult. Slit the intestine open and look at the inner surface. The ileum is lined with *villi,* microscopic projections into the lumen which give a velvety appearance to the gut's inner surface. The colon lacks villi, and its inner surface bears longitudinal ridges instead.

The mink has a pair of *anal glands* associated with the rectum. They produce evil-smelling musk and are usually removed during commercial preparation of dissection specimens. If they have not been removed, don't break them open.

### SUGGESTED READING

Aulerich, R. J., and D. R. Swindler, 1968. The dentition of the mink (*Mustela vison*). Jour. Mammalogy, 49: 488-494.

Gilbert, F. F., 1969. Analysis of basic vocalizations of the ranch mink. Jour. Mammalogy, 50:625-627.

Kainer, R. A., 1954. The gross anatomy of the digestive system of the mink. I. The headgut and the foregut. American Jour, of Veterinary Research, 15:82-90.

———. The gross anatomy of the digestive system of the mink II. The midgut and the hindgut. American Jour. of Veterinary Research, 15:91-97.

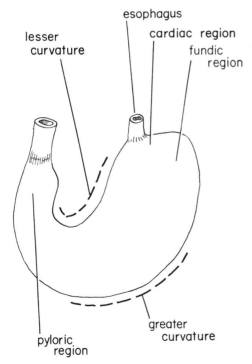

**FIGURE 27. The stomach.**

# Chapter 4
# Circulatory System

The circulatory system of the mink consists of *lymphatic ducts* and the *blood vascular system* (heart, arteries, veins, portal veins, and capillaries). Capillaries and lymphatics are difficult to study through gross dissection and are not covered in this manual. The arteries and veins of your specimen should be injected with colored latex—red for systemic arteries and blue for systemic veins. The hepatic portal system, if injected, should be yellow. If it is not injected, the vessels can be traced because the dark brown coagulated blood is visible through the thin walls. Use forceps and a blunt or flexible probe when tracing vessels.

Blood vessels are among the most variable structures of the vertebrate body. The branching patterns of the vessels in your specimen may differ greatly from those in the diagrams. Identify blood vessels primarily by the organs they supply and drain, and not by their branching pattern alone.

Arteries carry blood from the heart to capillary beds in either the lungs or the rest of the body. Arterial blood is under high pressure, and the walls of arteries are thick. Veins carry blood from capillary beds back to the heart. Venous blood is under low pressure, and the walls of veins are thin. Portal veins carry blood from one capillary bed to another without passing through the heart.

## THE HEART

Cut the pericardium and open the pericardial cavity. Note that the pericardium extends onto the great vessels connected to the heart and is reflected back on them and on the heart surface as the *epicardium,* or *visceral pericardium.* Cut the systemic aorta, the precava, the azygos vein, and the postcava (Figures 30, 31). Gently lift the heart outwards and cut the pulmonary arteries and veins as close to the lungs as possible. Try not to cut any of the nerves running alongside the pericardial cavity. The heart can then be removed from the body. Remove excess fat from the epicardium.

The *atria* (Figure 28) lie towards the right side of the chest. The *ventricles* are drawn to a point, the *apex,* on the left side. Identify left and right atria. Each atrium has an associated pouch, the *auricle,* which is visible externally. (The auricles are so named because of their fancied resemblance to ears if the heart is viewed as a face with the apex as the chin.) The atria are separated externally from the ventricles by the deep *coronary sulcus.* Right and left ventricles are separated externally by a shallow *interventricular sulcus* in the musculature. Identify the stumps of all blood vessels leading to and from the heart.

The heart musculature has its own blood supply, the *coronary arteries.* These arteries come off the systemic aorta and run in the coronary sulcus. Branches run from the sulcus to the atria and down the ventricles to the apex, supplying the muscular heart wall. The heart muscle capillaries are drained by a number of *cardiac veins.* Those draining the ventricular wall run from the apex toward the atria and empty into the *coronary sinus* on the dorsal surface of the heart. The coronary sinus empties into the right atrium.

Place the heart on the table and orient it with the ventral surface toward yourself and the apex pointing toward your right. Section the heart in the frontal plane so that your blade passes parallel to the table through the right ventricle and right atrium (Figure 29). Cut the wall of the pulmonary aorta. Remove the coagulated blood and latex from the heart and wash out the cavities. Be especially careful around the valves. Identify the chambers, valves, and blood vessels of the right side of the heart. Note the *chordae tendinae* and the *papillary muscles.* Note that the wall of the atrium is much thinner than the wall of the ventricle.

Turn the heart over and cut another frontal section on the dorsal side, parallel to the cut on the ventral

35

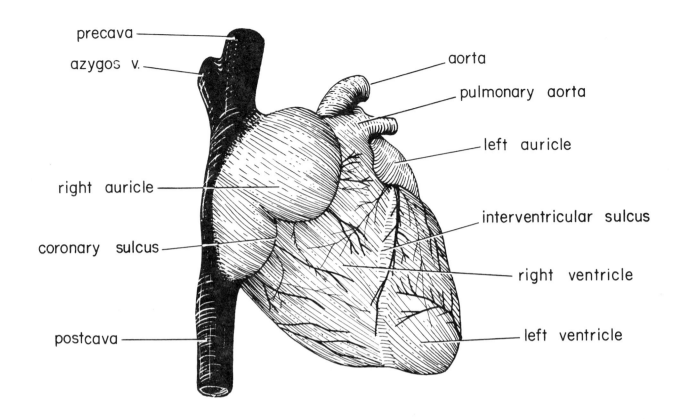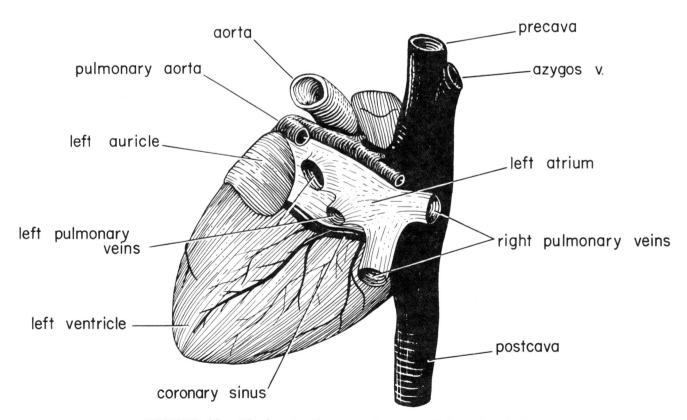

**FIGURE 28.** The heart. Above: ventral view. Below: dorsal view.

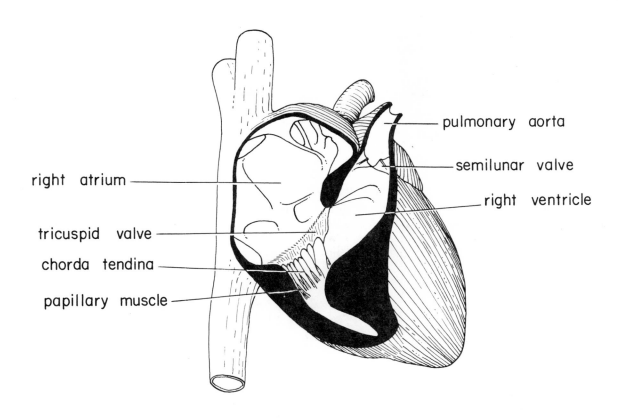
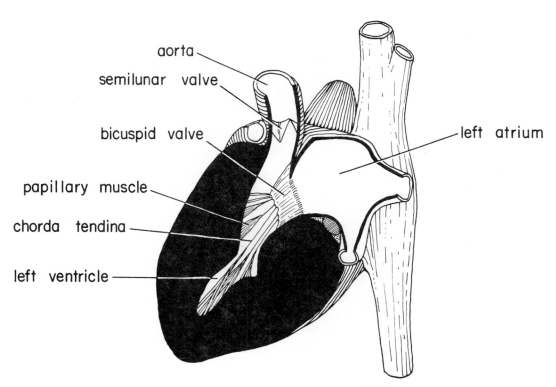

**FIGURE 29. Frontal sections through the heart.**
  Above: ventral view.
  Below: dorsal view.

side, but passing through the left atrium and ventricle. Extend the cut up the systemic aorta and clean out the cavities as before. Note that the wall of the left ventricle is thicker than that of the right, and the papillary muscles on the left side are much larger.

## ARTERIES AND VEINS

Clean the vessels in the thoracic cavity (Figures 30, 31). The systemic aorta curves dorsad as the *arch of the aorta* and then runs caudad as the *thoracic aorta*. Two major arterial trunks come off the arch of the aorta, the *brachiocephalic* and the *left subclavian*. The brachiocephalic gives off the *right internal thoracic artery* to the ventral chest wall and then divides into its three major branches, the *left* and *right common carotids* and the *right subclavian artery*. The branches of the subclavian should be dissected on the left side of the animal. The subclavian gives off three branches, the *costocervical artery* to the neck and anterior ribs, the *vertebral artery*, which penetrates the neck musculature and runs forward through the transverse foramina of the cervical vertebrae to the skull, and the *thyrocervical artery* to the neck. The subclavian then passes into the axilla and is known as the *axillary artery*. It gives off a large branch, the *subscapular artery*, and passes into the arm as the *brachial artery*. Trace it down the arm and through the entepicondylar foramen. In the forearm the brachial splits into the *radial* and *ulnar arteries*.

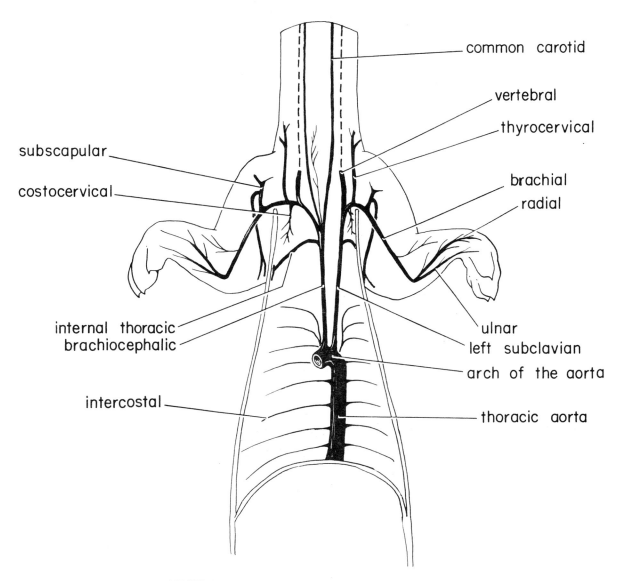

**FIGURE 30.** Arteries of the thorax and neck.

In the thorax the thoracic aorta gives off *intercostal arteries* to the musculature of the rib cage, *esophageal arteries* to the esophagus, and small *phrenic arteries* to the musculature of the diaphragm.

The systemic drainage of the front part of the body is collected in the *precava* (or *anterior vena cava*). It is formed by the fusion of the *right* and *left brachiocephalic veins*. Each of these receives blood from the *vertebral, internal jugular, external jugular,* and *subclavian veins*. The internal jugular vein runs alongside the common carotid artery, and the vertebral and subclavian veins are close to the arteries of the same names. In the thorax the precava also receives blood from a single *internal thoracic vein,* which drains from both sides of the ventral chest wall, and from the *azygos vein*. The azygos vein receives blood from the *intercostal veins* and is the homologue of the right posterior cardinal vein of lower vertebrates.

Trace the *external jugular vein* and its tributary veins on the surface of the right side of the neck and head. On the left side where the sternomastoid muscle has already been cut, locate the deeper vessels running alongside the trachea, the *internal jugular vein* and the *common carotid artery*. They run in a loose connective tissue sheath bound together with the vagus nerve. The internal jugular vein can be traced to its exit from the skull at the jugular foramen. It receives blood from sinuses in the skull and between the meninges of the brain.

Each common carotid artery gives off small branches to the esophagus and trachea and, just caudal to the origin of the digastric muscle, divides into the *internal*

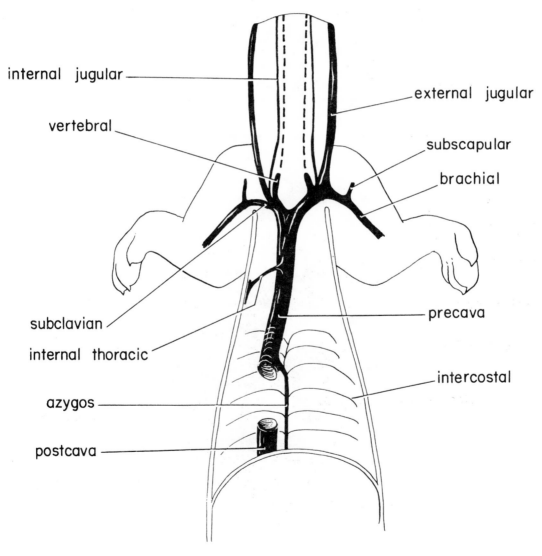

**FIGURE 31.** Veins of the thorax and neck.

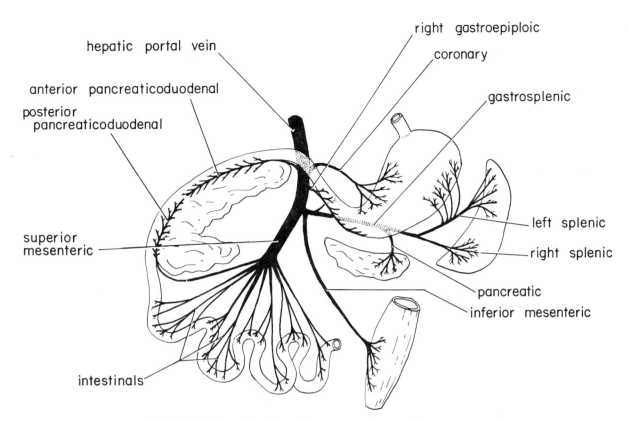

**FIGURE 32.** The hepatic portal vein and its branches.

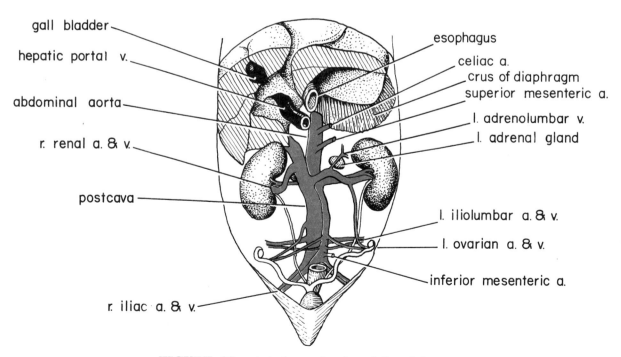

**FIGURE 33.** Arteries and veins of the abdomen.

and *external carotid arteries* (Figure 18). The external carotid artery continues in a ventral path and is easy to trace. The internal carotid, however, turns dorsad and runs alongside the medial edge of the tympanic bulla. It enters the skull through the posterior carotid foramen. The internal carotids and the vertebrals are the arterial suppliers of blood to the brain; the detailed distribution of their branches within the skull will be seen later during dissection of the brain. The external carotid branches into three major arteries. The *lingual artery* runs deep to the mylohyoid and between the styloglossus and hyoglossus to supply the tongue. The *maxillary* artery gives off several branches to salivary glands and penetrates the jaw musculature. It runs forward medial to the mandible and supplies the mandible, the jaw muscles, the orbit, and the snout. The *greater auricular artery* runs into the skin behind the pinna of the ear.

In the abdominal cavity, first expose and study the *hepatic portal vein* and its tributaries (Figure 32). This system of veins drains capillaries in the walls of the gut and carries the blood to the sinuses of the liver. The hepatic portal vein is formed by the junction of three major tributaries, the *superior* and *inferior mesenteric veins* and the *gastrosplenic vein*. The superior mesenteric vein is formed by a large number of *intestinal veins* from the small intestine and the *posterior pancreaticoduodenal vein,* which drains the wall of part of the duodenum and the right half of the right limb of the pancreas. The inferior mesenteric vein drains the wall of the lower part of the large intestine and the rectum. The gastrosplenic vein has a number of branches: the *pancreatic* from the left lobe of the pancreas, the *right splenic* from the spleen, the *left splenic* from the spleen and greater curvature of the stomach, and the small *middle gastroepiploic* (not labeled in Figure 32) from the pylorus. Cranial to the junction of the three major veins, the hepatic portal receives a number of smaller tributaries: the *right gastroepiploic* from the pylorus and greater curvature of the stomach, the *coronary* from the lesser curvature, and the *anterior pancreaticoduodenal* from the duodenum and the left half of the right limb of the pancreas. To expose the anterior and posterior pancreaticoduodenals, probe between the pancreas and the gut. The other branches of the hepatic portal vein are buried in the fat of the mesenteries. Try not to break any of the veins; when the blood drains from them they become invisible.

The liver has a dual blood supply. Blood in the hepatic portal vein is rich in nutrients freshly absorbed in the gut wall. Blood from the abdominal aorta (via a branch of the celiac artery) is rich in oxygen. The sinuses of the liver drain ultimately into the *hepatic veins,* which enter the postcava. The hepatic veins carry blood rich in waste materials and carbon dioxide. The hepatic veins may be found by cutting into the liver itself near the postcava.

The remaining vessels of the abdomen (Figure 33) are the *abdominal aorta* and its arterial branches, and the *postcava* (or *posterior vena cava*) and its tributaries. The abdominal aorta has three major branches to the gut, the *celiac* and the *superior* and *inferior mesenteric arteries*. The celiac artery splits into several branches which supply the liver, stomach, spleen, duodenum, and part of the pancreas. The superior mesenteric artery supplies most of the remainder of the intestines and the remainder of the pancreas. The inferior mesenteric artery supplies the lower part of the large intestine and the rectum.

The other branches of the abdominal aorta are associated with tributaries of the postcava; the *renal, adrenolumbar, iliolumbar, iliac,* and *caudal* vessels. The arteries to the gonads come off the abdominal aorta cranial to the iliolumbar branches; *ovarian arteries* in the female, *spermatic arteries* in the male. Venous return from the gonads enters the postcava on the right side and the renal vein on the left. The vessels of the hind limb are called the *external iliac artery* and *vein* inside the body cavity, and the *femoral artery* and *vein* in the thigh.

## SUGGESTED READING

Beddard, F. E., 1909. On some points in the structure of *Galidia elegans,* and on the postcaval vein in Carnivores. Proc. Zoological Society of London, 1909: 477-496.

# Chapter 5
# Urogenital System

Carefully remove the fat surrounding the kidneys and genital organs. Use forceps and a blunt probe. Save all the ducts and blood vessels. Expose the *kidneys*. They lie against the dorsal body wall and are covered by parietal peritoneum; they are said to be *retroperitoneal* in contrast with structures that hang down into the body cavity. The *adrenal glands* are small dark brown bodies lying in the fat medial to each kidney. The right adrenal gland lies dorsal to the right renal vein (Figure 33). Find and clean the *ureters* and trace them to their connections to the *urinary bladder*. The bladder is connected to the ventral body wall by a *suspensory ligament*. Urine passes from the kidneys to the bladder via the ureters and is stored there. The urine eventually passes from the bladder to the outside of the body through the *urethra*.

The kidney of the mink is bean-shaped, having a convex lateral border and an indentation, the *hilus*, medially (Figure 34). The ureter, renal artery, and renal vein enter the kidney at the hilus. Remove one kidney and slice it longitudinally in the frontal plane with a razor blade or sharp scalpel. Internally, two zones of tissue can be distinguished macroscopically: the outer granular *cortex,* and the inner striated *medulla*. The glomeruli and capsules of the kidney tubules are in the cortex, and the loops of the Henle and the collecting tubules are in the medulla. In the mink all collecting tubules converge at a single *papilla,* where the urine is emptied into a cavity, the *renal pelvis*. The renal pelvis is drained by the ureter.

## FEMALE REPRODUCTIVE TRACT

Expose the *ovaries, oviducts,* and *uterus* (Figure 35). Size and morphology of these structures vary with the reproductive state of the animal. If your mink is a fall-killed young female that has never born kits, the uterus will be thread-like and the ovaries and

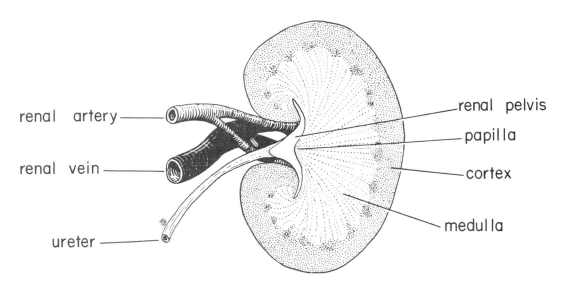

**FIGURE 34.** Kidney, frontal section.

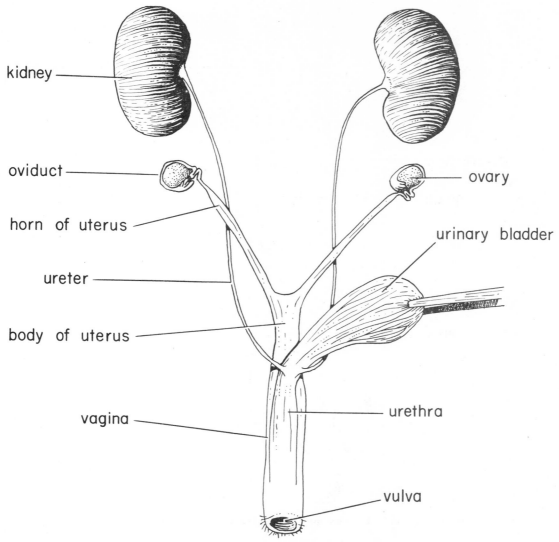

**FIGURE 35.** Urinary and genital tracts of the female (immature).

oviducts very small and difficult to study in detail. The uterus of the mink is *biocornuate,* having two *horns* (or *cornua*) which meet dorsal to the urinary bladder to form the *body* of the uterus. Each horn is supported by a sheet of mesentery called the *broad ligament.* Split the pelvis at the ischiopubic symphysis, bend back the two halves, and expose the lower part of the urinary and reproductive tracts. They end at the *vulva.* Probe the separate urinary passage, the *urethra,* leading from the bladder, and the reproductive passage, the *vagina,* leading from the uterus. Make a midventral incision in the vagina and uterus and try to find the distal limit of the body of the uterus, the *cervix.* The small *clitoris* may be found ventral to the vulva. It is the homologue of the male penis and has a small bone, the *os clitoridis,* the counterpart of the baculum.

The rectum may now be exposed and traced to its external opening, the *anus.*

## MALE REPRODUCTIVE TRACT

Find the *testicles* and clean them of fat (Figure 36). In the intact animal they are enclosed in a skin pouch, the *scrotum,* which is removed with the pelt. The tough sheath of the testicle is the *vaginal tunic,* an extension of the parietal peritoneum of the body cavity. Cut the tunic open and identify the *testis, epididymis,* and *vas deferens.* Sperm are produced in the testis, are stored in the epididymis, and eventually pass into the vas deferens. Trace the vas deferens to its entry into the abdominal cavity, over the ureter, and down the dorsal surface of the urinary bladder. Split the pelvis at the

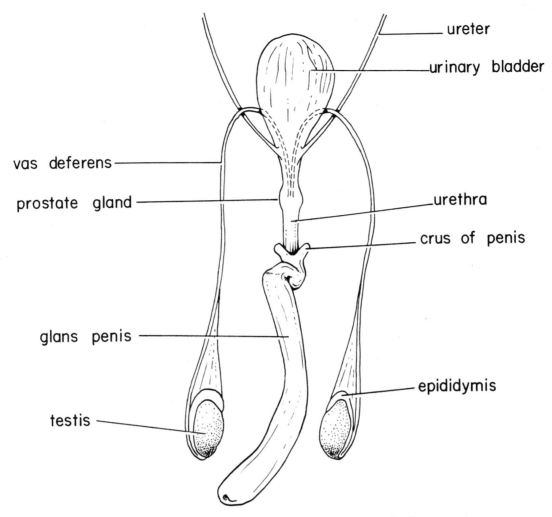

**FIGURE 36.** Urinary and genital tracts of the male (immature).

ischiopubic symphysis and bend back the two halves. At the base of the bladder is the *prostate gland,* not very distinct in the fall-killed young males. At the level of the prostate gland the vas deferens of each side opens into the urethra.

Some other accessory glands are associated with the urethra distal to the prostate gland. They cannot be seen in immature males, however.

The *penis* of the mink is relatively large. In the intact animal it points forward and is held against the ventral body wall in a sheath of skin. The penis is attached to the ischia by two *crura.* Distal to the crura are the paired, short *corpora cavernosa,* tough connective tissue cylinders which become engorged with blood during sexual arousal and cause erection of the penis. The corpora cavernosa are attached to the base of the *baculum,* or penis bone, which extends forward in the *glans* of the penis. The urethra passes between the crura of the penis and runs distally along the ventral surface of the baculum in the glans. Cut open the glans to expose the baculum. It has a sharp dorsal bend and a deep ventral groove in its distal half.

## REPRODUCTION IN THE MINK

During copulation the baculum protrudes from the glans penis, and its grooved ventral surface apparently presses against the cervix of the female. The semen is thought to flow from the end of the urethra in the glans along the groove of the baculum to the opening of the cervix. The semen passes through the body and horns of the uterus into the oviducts, and fertilization occurs there.

Female mink in North America usually come into heat and breed in March. Ovulation itself is in response to copulation, usually following it by 48 hours.

After fertilization the developing zygotes (or *blastocysts*) may remain in the upper regions of the uterine horns for many days, a phenomenon called *delayed implantation*. Eventually the blastocysts implant on the uterine lining (mucosa). Occurrence of delayed implantation in the mink doubtless accounts for the reported variation in pregnancy duration: from 40 to 75 days. The mean is approximately 51 days. Litter size varies from 1 to 17 kits. Most litters number 1, 2, 3, or 4, however.

## SUGGESTED READING

Enders, R. K., 1952. Reproduction in the mink (*Mustela vison.*) Proc. American Philosophical Society, 96:691-755.

Enders, R. K., and A. C. Enders, 1963. Morphology of the female reproductive tract during delayed implantation in the mink, pp. 129-139 *in* Delayed Implantation, A. C. Enders, ed., University of Chicago Press: Chicago, Illinois.

Gilbert, F. F., and E. D. Bailey, 1967. The effect of visual isolation on reproduction in the female ranch mink. Jour. Mammalogy, 48:113-118.

Hansson, A., 1947. The physiology of reproduction in mink (*Mustela vison* Schreb.) with special reference to delayed implantation. Acta Zoologica (Stockholm), 28:1-136.

MacLennan, R. R., and E. D. Bailey, 1972. Role of sexual experience in breeding behavior of male ranch mink. Jour. Mammalogy, 53:380-382.

Wright, P. L., 1963. Variations in reproductive cycles in North American mustelids, pp. 77-97 *in* Delayed Implantation, A. C. Enders, ed., University of Chicago Press: Chicago, Ill.

# Chapter 6
# Nervous System

## BRAIN AND CRANIAL NERVES

Remove the temporal muscles from the skull and clean the dorsal skull surface as far forward as the orbits. Remove the cervical muscles that attach on the occiput and clean the caudal surface of the skull down to the articulation with the atlas. Using bone cutters, clip into the lambdoidal crest at several places and remove the bone fragments. The bony *tentorium* (characteristic of the Order Carnivora) extends into the cranial cavity between the cerebrum and the cerebellum (Figure 24). Remove it in pieces. Working forward from the lambdoidal crest, pry off the bone from the dorsal and lateral surfaces of the brain. A bottle opener or the bottle-opening end of a "church-key" works better here than bone cutters, as it is less likely to damage the underlying brain. The tough connective tissue layer under the bone is the outermost *meninx* of the brain, the *dura mater*. It extends down between the cerebral hemispheres as the *falx cerebri*, and between the cerebrum and cerebellum as the *tentorium cerebelli*. The bony tentorium is an ossification in this part of the dura mater. Leave the dura mater intact until you are ready to remove the brain from the skull. Work forward removing bone from the skull until you reach the frontal sinuses, and downward until you reach the floor of the braincase. Remove also the occipital bones and expose the junction of the brain and spinal cord. Note the large venous sinuses in the dura mater and in the bones of the skull. These drain into the internal jugular vein via the jugular foramen.

When as much bone as possible has been removed, slit the dura mater from back to front and expose the brain. The innermost meninx, closely adhering to the surface of the brain itself, is the *pia mater*. Between the dura and pia layers is a loose network of fibers, the *arachnoid layer*. Cut the olfactory bulbs as far forward as possible and gently lift the front end of the brain. As the ventral surface of the brain is lifted, you will see the cranial nerves and blood vessels stretching downward to the floor of the braincase. Cut these as close to the skull (and as far from the brain) as possible. Most of the *pituitary gland* will remain behind in the skull in its bony housing, the *sella turcica*.

The most difficult part of this dissection is in the caudal region of the skull; parts of the cerebellum and the roots of cranial nerves VII through XII are usually damaged or destroyed as the brain is pulled out. Their remains can be studied in place in the skull after removal of the brain, however.

The arterial blood supply of the brain is ventral, via two pairs of arteries (Figure 38). The *vertebral arteries* pass through the foramen magnum from the transverse foramina of the cervical vertebrae. They lie on the ventral surface of the brainstem and join to form the median *basilar artery*. At the anterior border of the pons the basilar artery splits into two branches which pass around the pituitary stalk and meet anterior to it. These branches also receive blood from the *internal carotid arteries* which enter the skull through the posterior carotid foramina. The major arteries supplying the dorsal regions of the brain are the *posterior* and *anterior cerebellar arteries* and the *posterior, middle,* and *anterior cerebral arteries*.

Dorsally, the most prominent features of the brain are the *cerebral hemispheres*, separated by the deep *sagittal fissure*. The grooves in the cerebrum are called *sulci;* the bulges between them, *gyri*. A mammalian brain with its cerebral hemispheres grooved in this fashion is said to be *gyrencephalic*, in contrast to *lissencephalic* brains in which the cerebrum is smooth. Surface area and, hence, the total amount of cerebral cortex are proportionately greater in gyrencephalic brains. The *cerebellum* appears to lie directly behind the cerebrum; it also exhibits gyri and sulci. The cerebellum consists of a median *vermis* and right and left *lateral hemispheres*.

Ventrally the entire brainstem is visible, including the caudal *medulla oblongata,* the *pons* beneath the cerebellum, and the *hypothalamus*, which is the floor of the forebrain and which bears the pituitary gland. The *infundibulum* of the pituitary complex usually remains on the brain during removal of the brain from the skull. Anterior to the infundibulum the *optic nerves* enter the brain, after coming together at the *optic chiasma*. The cerebral surface is composed of the more dorsally located *neocortex* and the more ventral *paleocortex*. The paleocortical structures are the *olfactory tracts* and the *piriform lobes*. They are morphologically separated from neocortical structures by the *anterior* and *posterior rhinal fissures*. The paleocortex seems to be largely concerned with integration of information coming inward from the olfactory sensors. The olfactory nerve itself consists of a group of very short fibers originating in the *olfactory epithelium* of the nasal cavity and passing inward through a large number of small foramina in the bone of the anterior end of the braincase (the *cribriform plate*). They end in the *olfactory bulbs*, which lie directly behind the cribriform plate.

The roots of all twelve pairs of cranial nerves can be found either on the brain or in the skull. These nerves are more fully characterized in the appended table and in Figures 37 and 38.

After studying its external features, cut the brain in two in the midsagittal plane with a sharp scalpel or a razor blade. The structures on the cut surface are more visible if the brain is immersed in water. Left and right cerebral hemispheres are connected by major fiber tracts visible as the *corpus callosum* and the *fornix* (Figure 37).

The roof of the midbrain, the *tectum*, can now be seen between forebrain and hindbrain. On its dorsal surface are four small protuberances, the *corpora quadrigemina*. The anterior pair, called the *anterior colliculi*, are involved in control of the movements of the eyeballs. The posterior pair, the *posterior colliculi*, are centers of integration of acoustic information. The anterior colliculi are homologus to the optic lobes of the brains of lower vertebrates. In the sectioned *cerebellum* the central fiber tracts can be seen; these form the *arbor vitae* ("tree of ife").

The *fourth ventricle* is visible in the hindbrain beneath the cerebellum. It is part of a system of cavities (ventricles) and narrower passageways that carry cerebrospinal fluid within the brain. The mink's brain is so small that dissection of the anterior parts of this ventricular system is difficult.

## SPINAL CORD AND SPINAL NERVES

The spinal cord is the continuation of the central nervous system caudal to the foramen magnum. Paired spinal nerves arise from the cord and emerge through *intervertebral foramina*. Each spinal nerve is composed of two roots. The *dorsal root* in mammals carries only sensory fibers and has a ganglion. The *ventral root* in mammals carries only motor fibers. Beyond the junction of dorsal and ventral roots, the nerve divides into *dorsal* and *ventral rami*. Each ramus carries both sensory and motor fibers. The smaller dorsal ramus is distributed to epaxial musculature and the skin of the dorsal part of the body. The much larger *ventral ramus* is distributed to hypaxial and appendicular musculature and to the skin of the appendages and ventral part of the body.

By convention, the spinal nerves are named according to the vertebra near which they leave the vertebral column. The nerve between the skull and the atlas is called the first cervical, and the nerve between the seventh cervical vertebra and the first thoracic is called the eighth cervical. Caudal to this point each spinal nerve carries the name of the vertebra behind which it emerges, so that the nerve emerging between the sixth lumbar vertebra and the first sacral is called the sixth lumbar nerve.

The spinal cord itself stops growing before the vertebral column and trunk stop elongating, so that the cord does not extend into the lower lumbar and sacral regions. The lumbar and sacral nerves passing down the canal to exit at their appropriate intervertebral foramina are called collectively the *cauda equina* ("horse's tail"). The spinal cord is anchored to the base of the tail by a long filament called the *filum terminale*. The spinal cord is enlarged in the areas of motor outflow to the appendages, specifically in the caudal part of the cervical region and in the lumbar region. The ventral roots of spinal nerves in these areas are combined into complicated networks called the *brachial plexus* (associated with the pectoral limb) and the *lumbosacral plexus* (associated with the pelvic limb).

To expose the brachial plexus, remove the pectoral muscles on the right side of the body, noting the nerve that supplies them. Clean the fat and fascia from the nerves, and remove the blood vessels in the right axilla. Clean and identify the nerves distally in the arm and work back toward their origins in the brachial plexus (Figure 39). The *median nerve* runs through the entepicondylar foramen together with the brachial artery. The *ulnar nerve* passes behind the medial epicondyle and beside the olecranon process to enter the foramen. The median and ulnar nerves supply the flexor muscles in the forearm and forepaw. The *musculocutaneous nerve* supplies the biceps brachii and brachialis muscles in the arm. The large *radial nerve* enters the triceps muscle between the medial and long heads. It supplies the triceps, anconeus, and the extensor muscles in

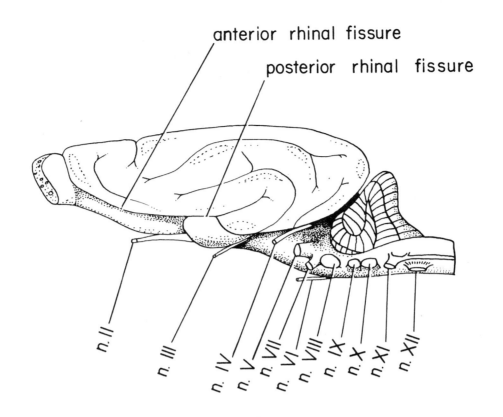
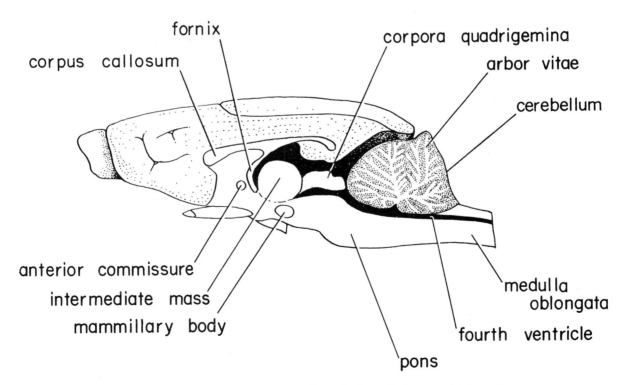

**FIGURE 37.** The brain. Above: lateral view. Below: sagittal section.

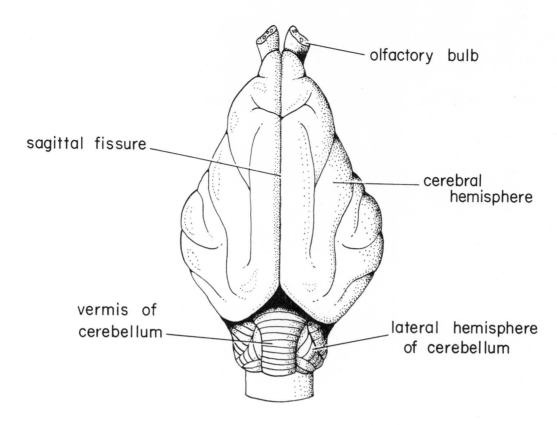
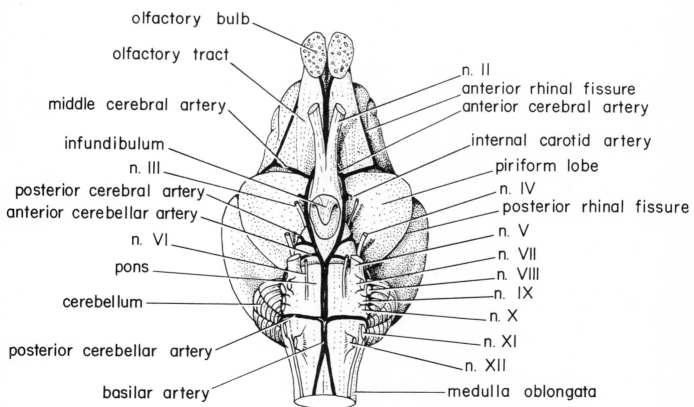

**FIGURE 38.** The brain. Above: dorsal view. Below: ventral view.

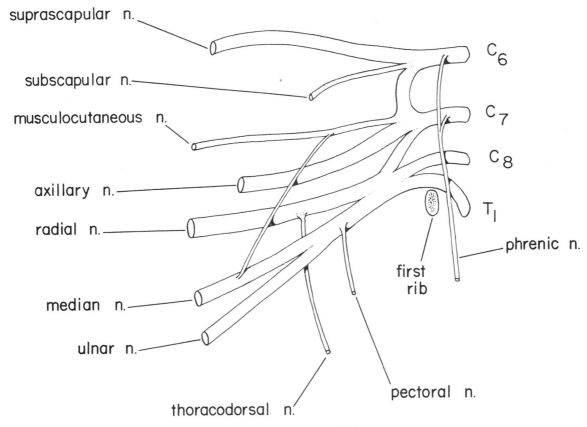

**FIGURE 39. The brachial plexus.**

the forearm. The *axillary nerve* passes through the shoulder between the teres major and subscapularis muscles and supplies the deltoids. The *thoracodorsal nerve* comes off the radial and supplies the latissimus dorsi. The largest nerve in the anterior end of the plexus is the *suprascapular nerve,* which runs onto the dorsal surface of the scapula and supplies the supraspinatus and infraspinatus. The small *subscapular nerve* supplies the subscapularis. Many of these nerves also carry sensory impulses inward from receptors in the skin.

The slender *phrenic nerve* originates from the sixth and seventh cervical nerves and passes caudally in the thorax. It lies in the fat in the middle of the cavity, passes ventrally over the root of the lung, and then diverges laterally to innervate the muscular diaphragm.

The brachial plexus of the mink is formed by ventral rami of the last three cervical nerves (six through eight) and the first thoracic.

The lumbosacral plexus is formed by ventral rami of the fourth through sixth lumbar nerves and the first two sacral nerves (Figure 40). It is more difficult to dissect than the brachial plexus, but the method is the same: identify the major nerves peripherally in the hind limb and trace them inward to their origins in the plexus. It will be necessary to cut and remove parts of the psoas minor and iliopsoas muscles. Try to leave the circulatory, digestive, and urogenital systems intact insofar as possible. The largest nerve is the *sciatic nerve* (or *ischiadic nerve*). In the thigh it lies deep to the biceps femoris and the tenuissimus (Figure 20). It is actually a compound of two separate nerves; the *tibial nerve* which innervates the hamstring muscles of the thigh, the gastrocnemius and associated muscles of the leg, and the flexor muscles of the hindpaw, and the *common peroneal nerve* which supplies the tenuissimus and the muscles on the cranial surface of the leg. At the point of emergence from the pelvis the sciatic gives off a small branch, the *gluteal nerve* to the gluteal muscles, tensor fascia lata, and the femorococcygeus. The *femoral nerve* emerges from the pelvis ventrally and supplies the sartorius and the quadriceps femoris. The *obturator nerve* is best found by separating the two halves of the pelvis ventrally and looking at the medial surface of the innominate bone. The nerve passes through the obturator foramen to innervate the

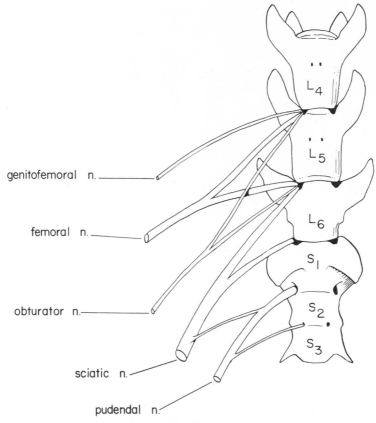

**FIGURE 40. The lumbosacral plexus.**

adductor muscles, the external obturator, and the gracilis. The *pudendal nerve* runs caudally within the pelvic cavity and innervates the external genitalia.

## AUTONOMIC NERVOUS SYSTEM

The autonomic nervous system of mammals can be divided into two functional and morphological parts. The *sympathetic part* consists of a set of fibers leaving the spinal nerves in the thoracic and lumbar regions. They enter a chain of ganglia lying close to the vertebral column. The ganglion chain extends into the cervical region, but is fused there with the vagus nerve. From the ganglion chain, nerves run to numerous visceral structures including the salivary glands, intrinsic muscles of the eye, the heart, lungs, digestive tract, adrenal glands, and the reproductive organs.

The *parasympathetic part* consists of two separate groups of fibers. The first group leaves the brain along with cranial nerves III, VII, IX, and X. The second group leaves the spinal cord along with several sacral spinal nerves. The parasympathetic ganglia are located far from the vertebral column, close to the structures being innervated. Each visceral structure innervated by sympathetic fibers receives parasympathetic innervation also, and the actions of the two parts of the system are opposite to one another.

Dissection of the autonomic nervous system of the mink is extremely difficult. You should be able to find the sympathetic chain of ganglia in the thorax and the combined vagosympathetic trunk in the neck. Finding the other nerves and ganglia in the fat of the abdominal cavity is possible if you have a lot of time, skill, and patience.

### SUGGESTED READING

Pilleri, G., 1960. Das Gehirn von *Mustela vison* und *Mephitis mephitis* (Carnivora, Mustelidae). Review Suisse de Zoologie, 67:141-158.

Radinsky, L., 1968. Evolution of somatic sensory specialization in otter brains. Jour. Comparative Neurology, 134:495-506.

Thiede, U., 1966. Zur Evolution von Hirneigenschaften mitteleuropäischer und südamerikanischer Musteliden. Zeitschdift für Zoologische Systematik und Evolutionsforschung, 4:318-377.

## THE CRANIAL NERVES

| NAME | SUPERFICIAL ORIGIN ON BRAIN | FORAMEN OF EXIT FROM OR ENTRANCE TO CRANIAL CAVITY | COMPOSITION AND DISTRIBUTION |
|---|---|---|---|
| I. Olfactory | Olfactory bulb | Cribriform plate | Sensory from olfactory epithelium |
| II. Optic | Thalamus | Optic foramen | Sensory from retina of eye |
| III. Oculomotor | Cerebral peduncles | Combined foramen rotundum and orbital fissure | Motor to superior, inferior, medial recti; inferior oblique muscles of eyeball |
| IV. Trochlear | Dorsal surface of midbrain | Combined foramen rotundum and orbital fissure | Motor to superior oblique muscle of eyeball |
| V. Trigeminal | | | |
| V₁ Profundus | Pons | Combined foramen rotundum and orbital fissure | Sensory from skin of head and rostrum |
| V₂ Maxillary | Pons | Combined foramen rotundum and orbital fissure | Sensory from skin of face and upper teeth |
| V₃ Mandibular | Pons | Foramen ovale | Motor to jaw muscles; sensory from mandible |
| VI. Abducens | Anterior medulla | Combined foramen rotundum and orbital fissure | Motor to lateral rectus muscle of eyeball |
| VII. Facial | Anterior medulla | Stylomastoid foramen | Motor to superficial and deep facial muscles; taste sensation |
| VIII. Acoustic | Anterior medulla | Internal acoustic meatus | Sensory from inner ear |
| IX. Glossopharyngeal | Anterior medulla | Jugular foramen | Sensory and motor; pharynx, base of tongue |
| X. Vagus | Anterior medulla | Jugular foramen | Some sensory components; motor to laryngeal muscles; parasympathetic distribution to heart, lungs, gut, etc. |
| XI. Accessory | Medulla and spinal cord | Jugular foramen | Motor to sternomastoid, cleidomastoid, trapezius |
| XII. Hypoglossal | Medulla | Hypoglossal foramen | Motor to intrinsic and extrinsic muscles of tongue |

# Chapter 7
# Sense Organs

## THE EYE

The eyes of the mink are quite small, and you may wish to use supplementary material from another mammal, such as the sheep. The eyeball is moved in the orbit by a group of six extrinsic muscles, the *superior, inferior, medial,* and *lateral recti,* and the *superior* and *inferior obliques.* Cut the eyeball in half with a razor blade or sharp scalpel. The tough outer coat of the eyeball itself is the *sclera* (Figure 41). Inside this lies the darkly pigmented *choroid layer.* Inside the choroid layer is the *retina.* Light enters the eye through the *cornea* (an extension of the sclera over the front of the eyeball) and passes through the *lens* into the *vitreous chamber.* Light passes through the retina, exciting the sensory cells there, and is absorbed in the choroid layer. In mammals the lens is suspended from the ring-like *ciliary body* by suspensory ligaments. Ciliary muscles, acting through the ligaments, change the shape of the lens and allow the eye to accommodate. Intrinsic muscles in the *iris* control the size of the pupil and thus control the amount of light that enters the eye. The iris functions very much like the diaphragm of a camera, opening wide at dusk to allow the maximum amount of light to enter the eye, and closing down in bright light.

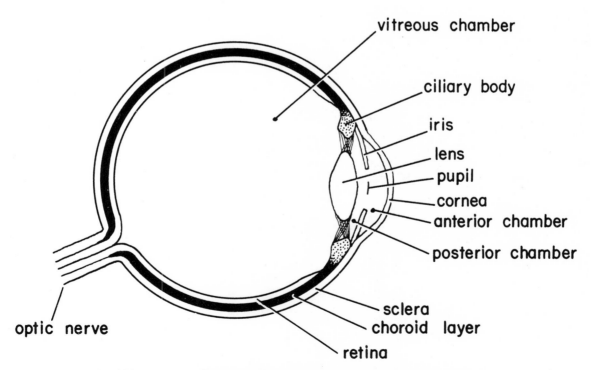

**FIGURE 41. Diagrammatic section of the eyeball of a mammal.**

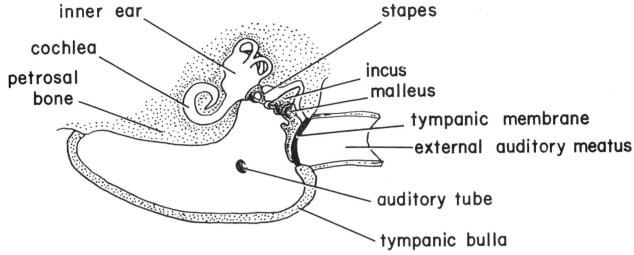

**FIGURE 42.** Diagrammatic cross section through the ear region of the mink.

## THE EAR

Dissection of the ear in the mink is very difficult. If you have the time and equipment, remove the mandible, musculature and connective tissue from the ventral surface of the skull, exposing the tympanic bulla. Remove the connective tissue of the external auditory canal until you reach the external auditory meatus of the skull. Using a heavy pair of scissors and working under a dissecting microscope, chip away at the bone of the bulla until you see the *tympanic membrane* (eardrum), Figure 42. The *malleus* is attached to this membrane and articulates with the *incus,* which lies in a dorsal recess in the petrosal bone. After removal of the malleus and incus you should see the stirrup-shaped *stapes,* which attaches to a fenestra in the bony housing of the membranous labyrinth of the inner ear. Note that the inner surface of the tympanic bulla is complicated by numerous struts and partitions.

The mammalian ear is divided into three regions: the *outer ear* (the pinna and the external auditory canal, to the tympanic membrane), the *middle ear* (the chamber containing the malleus, incus, and stapes, and communicating with the pharynx by the auditory tube), and the *inner ear* (the membranous labyrinth, enclosed within the petrosal bone). The reptilian ancestors of mammals had a middle ear cavity with a single auditory ossicle, the stapes. During the evolution of mammals from reptiles the bones of the old reptilian jaw joint (the quadrate and articular) were transformed into auditory ossicles (the incus and malleus, respectively), producing an auditory apparatus with far greater sensitivity than the reptilian one.

### SUGGESTED READING

Hopson, J. A., 1966. The origin of the mammalian middle ear. American Zoologist, 6:437-450.

Segall, W., 1943. The auditory region of the arctoid carnivores. Zool. Series of Field Museum of Natural History, 29:33-59.

Walls, G. L. The vertebrate eye and its adaptive radiation. Cranbrook Institute of Science, Bulletin No. 19, pp. i-xiv, 1-785.

# Chapter 8
# Skin

Though the skin of your animal has been removed, it is worthwhile to consider briefly the major aspects of structure and function of this organ. The skin is formed by the superficial *epidermis* and the deeper *dermis* (Figure 43). The thinner epidermis is itself multi-layered, the outer *stratum corneum* consisting of lifeless keratinized cells which are progressively shed, and the inner *stratum germinativum,* where new cells of the stratum corneum are produced. The thicker dermis is formed by connective tissue, with supplies of nerves and blood vessels. Its outer surface indents the underside of the epidermis as a series of *dermal papillae.*

The skin forms a tough complete covering for the whole body, prevents body fluids from being lost, and protects the body from injury by microorganisms or physical damage. The furred skin of the mink is also

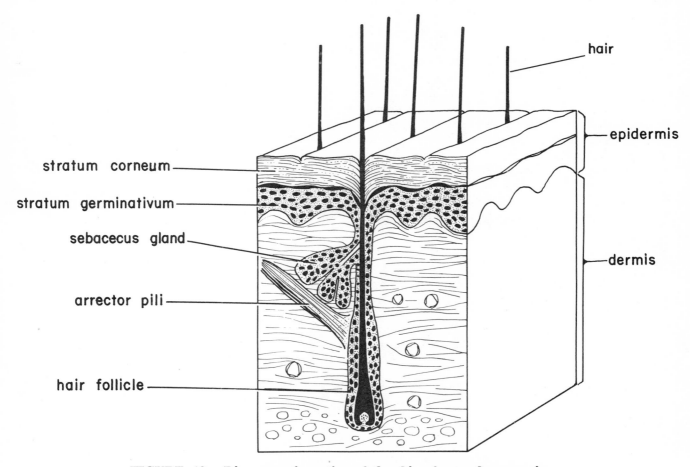

**FIGURE 43. Diagrammatic section of the skin of a newborn carnivore.**

crucial in the animal's thermoregulation. Air spaces trapped in the fur impede the outflow of heat from the body to a cold environment. In some mammals, including humans and horses, the skin is also important in thermoregulation in hot environments. Sweat, produced by *sudoriferous glands,* evaporates from the skin and cools the body. Carnivores such as the mink have sudoriferous glands mainly on the soles of the paws and must lose heat by panting. The skin and its derivatives are also important in sensing various features of the environment.

Hairs and glands project deep into the dermis, though their embryonic origin is epidermal. Each hair is rooted in an epidermal tube called a *follicle* and grows from a *papilla* at its base. Most of the hairs on the body are of two sorts, the short wooly *underfur* and the longer sparser *guard hairs*. Associated with each hair is a bundle of smooth muscle, the *arrector pili,* and one or more *sebaceous glands*. The arrector pili moves the hair when the fur is "stood on end." The sebaceous glands produce oil, which lubricates the hair and the outer layer of the epidermis. The oily guard hairs of a healthy mink are effective water repellants. Fur in the mink is shed and renewed twice yearly, in spring and fall. The thick winter coat when fully developed is said to be *prime*. Mink used in the fur trade are skinned when prime.

In a newborn carnivore the hair follicles are isolated (Figure 43), but in adults the follicles occur in bundles. A single guard hair emerges from a deep follicle in the center of each bundle. The other hairs in the bundle form the underfur and emerge from shallower follicles. Fur in mustelids slopes uniformly caudalwards on the body. In other carnivores, such as dogs and wolves, hair slope varies in a complex fashion over the body.

Longer stiffer sensory hairs are called *vibrissae* and are found on the head. The longest are the *mystacial vibrissae* (or "whiskers") extending outwards from the muzzle. The *genal vibrissae* on the cheeks, the *supraorbital vibrissae* over the eyes, and the *mental vibrissae* on the chin are shorter. Vibrissae are also called *sinus hairs* because the base of each is enclosed in a blood-filled sinus in the dermis. Pressure sensors in the walls of the sinus are activated when the vibrissa contacts some object.

Other sorts of glands in addition to the sebaceous and sudoriferous types are developed in the skin. A pair of *anal glands* produce musk. Three pairs of *mammary glands* are developed in the female. Other derivatives of the skin are the *claws* on the front and hind paws.

To the underside of the skin are attached sheets of striated muscles. In the head and neck these muscles belong to the *superficial facial group* and are innervated by cranial nerve VII (the facial nerve). In the trunk region, the dermal muscle sheet is the *panniculus carnosus,* which is phylogenetically an appendicular muscle and is innervated by pectoral nerves from the brachial plexus. These dermal striated muscles are not related to the smooth muscles of the hairs, the arrectores pilorum. The area beneath the skin is filled with loose connective tissue and heavy deposits of *subcutaneous fat.*

## SUGGESTED READING

Hildebrand, M., 1952. The integument in Canidae. Jour. Mammalogy, 33:419-428.

Huber, E., 1930. Evolution of facial musculature and cutaneous field of trigeminus. Quarterly Review of Biology, 5:133-188, 389-437.

Lovell, J. E., and R. Getty, 1964. The Integument, pp. 875-888 in Anatomy of the Dog, M. E. Miller, G. C. Christensen, and H. E. Evans, Saunders: Philadelphia and London.

Patrizi, G., and B. L. Munger, 1966. The ultrastructure and innervation of rat vibrissae. Jour. Comparative Neurol., 126:423-436.